MEN & MACHINES OF THE AUSTRALIAN FLYING CORPS 1914–19

Charles Schaedel
Illustrated by the Author

KOOKABURRA TECHNICAL PUBLICATIONS PTY LTD
81 POTTER ST, DANDENONG 3175, VICTORIA, AUSTRALIA

"As it was in the beginning"—Lt Richard Williams flies a Bristol Boxkite at Point Cook, beginning a career which culminated in his becoming Air Marshal Sir Richard Williams, KBE, CB, DSO.

INTRODUCTION

The Royal Australian Air Force Golden Jubilee year of 1971 has focused attention on a dedicated band of aviation pioneers whose valiant deeds during World War I established the traditions and laid the foundations of the RAAF. These were the men of the Australian Flying Corps.

The AFC was a prime example of national self-awareness, at that time unique among the air services of other British Dominions in World War I. Its members came from all walks of life, and the consequent cross-section of character and skills produced a highly adaptable and very efficient fighting force which earned praise and respect in every theatre in which it served.

Unfortunately, a brief description such as this makes it impossible to give credit to all who played such an important part in achieving final success, be they airmen or ground staff. The names mentioned are used to illustrate certain aspects of the air war in order to give as complete a picture as possible of the work performed and the dangers faced by the men of the AFC, rather than to highlight the deeds of individuals. Similarly, it is regretted that a work devoted to a national force must necessarily preclude mention of the many Australian airmen who served with great distinction in other Allied services, in some cases with their nationality unrecognised in the history books.

This book is intended as a tribute to Australians who were associated, either in the air or on the ground, with the air battles of World War I. After that terrible conflict, many of the young men of the AFC returned to Australia determined to use their professional abilities to lasting effect and pioneered civil aviation in Australia. Indeed, some of them pioneered not only domestic, but world air routes, and helped to lay the foundations of air transport as we know it today.

Charles Schaedel
Adelaide, 1972

"—and at the end!" A Bristol Fighter of No 3 Squadron AFC portrays the strength and aggressiveness which typified the Australian air combat units in World War I.

CONTENTS

ACKNOWLEDGEMENTS

The author gratefully acknowledges the many people who made this work possible by providing photographs, information and material assistance, particularly the following (listed in alphabetical order): Australian War Memorial, A. R. Betteridge, N. R. Burnell, N. Clutterbuck, E. A. Collins, W/O J. Cramphorne (RAAF Museum, Point Cook), J. R. Gordon, Air Marshal Sir George Jones, KBE, CB, DFC, Mrs F. J. Mart, Mrs R. O. C. Matthews, R. H. Mitchell, K. A. Merrick, G. Pentland, E. F. Pflaum, J. C. Sands, Lady V. White, R. Whitfield (Xenon Photo Services), Air Marshal Sir Richard Williams, KBE, CB, DSO, and last but not least, his own family for their patient forbearance. Apologies are extended to anyone inadvertantly overlooked. The author also wishes to accept full responsibility for the poor quality of a small number of photographs which were deemed worthy of inclusion on account of their unique historical value.

KOOKABURRA TECHNICAL PUBLICATIONS PTY LTD

81 POTTER ST, DANDENONG 3175, VICTORIA, AUSTRALIA

TYPE SET BY MEREDITH TRADE LINO PTY LTD, MELBOURNE

PRINTED BY TOPPAN PRINTING CO LTD, TOKYO, JAPAN

Fledgling wings! Central Flying School aircraft at Point Cook in 1915. Bristol Boxkite in foreground; BE2a and Bleriot monoplane behind.

This Deperdussin monoplane was used for taxying practice until one adventurous pilot vowed he would fly it. The result is shown here, and the repairs to its broken back are evident on the still existent aircraft today.

The Bleriot monoplane was strictly functional and its simple construction is well illustrated here, with Air Mechanic R. H. Mitchell in the cockpit.

Air Mechanic Mitchell demonstrates starting procedure on a BE2a at Point Cook. Mitchell later escaped with relatively minor injuries in England when drawn into the high-revving propeller of a Sopwith Camel after the pilot failed to throttle back on starting.

CHAPTER 1 POINT COOK AND MESOPOTAMIA

On the outbreak of World War I on 4th August 1914, Australia had just embarked on the formation of a military aviation branch of its armed services. During 1913 land had been acquired at Point Cook (at this time spelt Cooke) near Melbourne for use as a training base for military pilots who would comprise the Australian Flying Corps, and arrangements were in hand to begin the first course when hostilities commenced. Thus the Central Flying School was already on a war footing when the first four candidates began their instruction under the watchful eyes of Lt Eric Harrison and Lt Henry Petre. Harrison, an Australian, and Petre, an Englishman, had learnt to fly in England before being selected to train pilots for the AFC, and to carry out this task, they were initially provided with five aircraft—two BE2a biplanes, two Deperdussin monoplanes and one Bristol Boxkite. The first four pupils, with the now familiar names of Capt T. W. White and Lts R. Williams, G. P. Merz and D. T. W. Manwell were officers voluntarily seconded from their army units. In view of all the problems associated with the brand-new venture, it is a credit to everybody concerned that they all qualified as pilots before the end of the year. The first was Lt Williams, setting him on the path which culminated in his becoming Air Marshal Sir Richard Williams, KBE, CB, DSO.

Point Cook continued to provide trained air and ground crews throughout the war, a total of eight classes being completed, but compared to overseas training units the flying equipment remained primitive. At one stage this included a Bleriot monoplane similar to the one in which the first crossing of the English Channel was made in 1909. A couple of Caudron pusher biplanes improved the situation when added to the fleet, and another type which was recognised as ideal for the training role was the Maurice Farman Shorthorn, but the little Deperdussins resisted strongly any attempts to get them off the ground.

As the school grew, Harrison and Petre were assisted at various times by J. R. Duigan (credited with having made the first powered flight in an Australian-built machine in 1910), Lt W. H. Treloar, Capt Delfosse Badgery and Sgt Charles Marduel, an experienced French pilot. Actually, Point Cook served as an Elementary Flying Training School, to use a later designation, with pupils receiving their final training at units in England. Less than 100 passed through the Australian station, but this number was boosted by other pupils trained by W. Stutt, a UK-trained Australian pilot who ran a flying school at Richmond, New South Wales, using American Curtiss Jennys. Fittingly enough, his school was under the control of the New South Wales Education Department, and it did a very useful job in supplying pilots for final instruction in the Australian training units established in England in the final year of the war. These units, comprising Nos 5, 6, 7 and 8 Training Squadrons, formed a training wing that was wholly Australian, an independent set-up unique among the Dominions, and which did much to assist the establishment of the post-war Air Force. This wing was commanded by Lt-Col Oswald Watt, an Aus-

The 50 hp Gnome rotary en-gine of a Bristol Boxkite examined by Air Mechanic A R. Betteridge.

tralian who learnt to fly in 1911 and whose subsequent experiences included pre-war flying in Egypt, active service with the French Air Service, more desert flying with No 1 Squadron, AFC, and finally command of No 2 Squadron, AFC, in France.

Long before this period, however, the Australian Flying Corps took part in its first active service operation when a small party was ordered to accompany an expeditionary force to New Guinea toward the end of 1914 with the object of securing the German settlements in this area. As a show of tactical air strength the operation did not even get off the ground, for the occupying forces met no resistance and the two aircraft, one of them a BE2a from Point Cook and the other a Farman seaplane, remained in their crates until the return to Australia. The two pilots chosen for this mission were Lt Harrison, the instructor, and Lt Merz, whose hour of fame was near at hand.

Early in February 1915 the Australian Government received a request from the Indian Government for the supply of an air force unit to aid its operations against the Turks in Mesopotamia, a request which seems strange in view of the size of the AFC at the time unless some exaggerated statements had been bandied about. Nevertheless, it was agreed that Australia would supply the force if India could supply the aircraft, five being requested (three were eventually supplied, consisting of two Maurice Farman Shorthorns and a Maurice Farman Longhorn). Of the seven qualified pilots in Australia who were under military jurisdiction, four were detailed as the flying section of the unit, with Petre in command of his erstwhile pupils White and Merz, together with Treloar who had learnt to fly in England. The rest of his command consisted of 41 other ranks, very few with any experience of aircraft but selected because of their abilities in allied fields. Together they formed what became known as the Mesopotamian Half-Flight.

Hasty training and gathering of materials finally resulted in the main force leaving Australia toward the end of April 1915 in the charge of White and Treloar, Petre having gone ahead the previous week. Merz remained at Point Cook for a short period to assist with the training programme. Their destination at Basra was reached on 26th May when they were joined by two British airmen from the Indian Army and a New Zealander. They were greeted with the news that they were just in time to take part in a planned advance up the Tigris River. Any visions that the Australians may have had about glorious adventure were quickly dispelled in the following days. It soon became evident that they would have to fight a constant battle not only against the enemy but against the problems of heat, sand, poor equipment, lack of replacements and, from

Lt R. Williams prepares to fly a BE2a at Point Cook. Right. A Maurice Farman Hydroplane at Point Cook in 1915. Possibly this machine and the BE2a at left were the aircraft sent to New Guinea in 1914.

Left. A later introduction for training purposes was the Maurice Farman Shorthorn, an example of which is shown here bearing CFS number 15. Right. Bristol Boxkite in flight at Point Cook. The amount of breeze indicated by the windsock is probably the maximum permissible for this aircraft to be flown!

the pilots' point of view, totally unsuitable aircraft. These, if they could be coaxed into the hot desert air, inspired no confidence that they would remain there. These depressing conditions also faced Lt Merz when he finally arrived in mid-June, although the addition of two Caudrons a few weeks later improved the situation slightly. Still highly unsuitable for the conditions prevailing, the Caudrons represented a slight improvement over the Farmans. Any improvement, however, was a good thing if it gave that little bit of extra confidence necessary for the long flights over the marshes of the Euphrates.

It was on one of these flights that the Australian Flying Corps suffered its first casualty on flying operations. The two Caudrons had completed a reconnaissance mission and were returning to Basra on 30th July when one of them, flown by Capt Reilly of the Indian Army, was forced down by engine trouble, fortunately among relatively friendly Arabs. The other machine, flown by Merz with the New Zealander Lt Burn as his passenger, continued toward base but was also forced to land, in this case among hostile and well-armed Arabs. The fliers, armed only with revolvers, were overcome after they had put up a gallant battle. Although the two airmen were never seen again, the story of their running fight was told to the authorities by friendly Arabs, the account being confirmed a few days later when the Caudron was found, virtually destroyed by the attackers.

The following month saw the establishment of the Half-Flight increased by the addition of four Martinsyde scouts and three seaplanes, at which stage it became No 30 Squadron, Royal Flying Corps, but the performance of the new aircraft types was such that the existing problems were only multiplied. Further misfortune befell the Australian contingent on 16th September, when Treloar's Caudron came down behind the Turkish lines and he and Capt Atkins of the Indian Army were captured by Arabs and handed over to the Turks to suffer the terrible conditions of the prisoner-of-war confines. This and other losses left only a Shorthorn and a Martinsyde to cover the attack on Kut on 27th September, 1915, although three seaplanes were available for co-operation with the artillery. After the successful occupation of Kut a landing ground was established, providing a better chance to maintain an effective force. From this base the British and Australian pilots carried out reconnaissance flights up the Tigris; long and dangerous flights in their slow, unarmed and unreliable machines and with full knowledge of the reception awaiting them at the hands of the enemy if they were forced down.

White was forced to taxi his Shorthorn along the ground for 20 miles on one occasion when the engine failed to provide sufficient power to keep the machine in the air. Luck was with him and despite the fact that at one stage he passed in close proximity to a heavily defended Turkish position, he reached base safely. In November, White also figured in what must have been one of the first desert air rescues (later in the same month Squadron Commander R. Bell Davies of the Royal Naval Air Service was awarded the Victoria Cross for rescuing a downed pilot from under the noses of enemy troops in the Dardanelles). White's feat consisted of landing in the face of rifle fire from hostile Arabs to pick up Maj-Gen G. V. Kemball, the Chief of General Staff in Mesopotamia, whose seaplane had been forced down with engine trouble. Fortunately, an Indian cavalry patrol was near at

The first tragedy of the AFC war was the loss of Lts Merz and Burn in Mesopotamia. Their Caudron was eventually salvaged and brought back to base by barge and motor transport.

A French Voisin aircraft operated by 30 Squadron RFC in Mesopotamia. The AFC First Half-Flight became a part of this unit.

hand to bring home the General's pilot, otherwise his extra weight would have necessitated another long ground run by White. Shortly after this, on 13th November, his luck ran out on what was virtually a voluntary suicide mission to cut the telegraph wires near the enemy-held city of Baghdad. This task involved a round trip of 120 miles carrying extra load in the form of guncotton, petrol and oil, and was possible only if the aircraft performed perfectly and no head winds were encountered. Apart from these factors, the only other worry was the large number of Turkish troops in the region of the target! Choosing an area that offered a fair chance of success without too much interference, White landed as close as possible to the wires, so close in fact that he badly damaged one wing against a telegraph pole, but making it possible for his companion Yeats-Brown to blow up the lines even though under fire from Arabs and Turks. They then attempted to taxi clear of the enemy but were unsuccessful and were taken prisoner.

This loss left Capt Petre and one Shorthorn as the sole Australian aerial component of 30 Squadron, which still continued to operate during the following days until the Allied troops were finally halted. Thrown back and beseiged in Kut by the Turks, they succeeded in holding out until forced to surrender on 29th April 1916. Petre was fortunate to escape this fate as the remaining aircraft were ordered out of Kut in December, and he was sent to Egypt with a number of the Australian mechanics to await the arrival of the first unit of the Australian Flying Corps from Australia. Nine other Australian mechanics of the ill-fated Half-Flight remained trapped in Kut, and of these only two, Sgt Sloss and A/M Hudson, together with their previously captured officers Treloar and White, survived their eventual captivity.

Thus ended a disastrous campaign of which it can only be said that in this pitiless and isolated area the embryo Australian Flying Corps fought its own Gallipoli, and established the traditions which were to make it an outstanding force in the years to follow.

The pilots of the First Half-Flight AFC at Basra in 1915. From left are Lt W. H. Treloar, Capt T. W. White, Capt H. A. Petre, and Lt G. P. Merz with their Indian batman seated in front. Right. Comprehensive diaries maintained by Capt T. W. White during years of captivity and privation were successfully hidden in his boots as shown. These notes formed the basis of his classic book on escape entitled "Guests Of The Unspeakable."

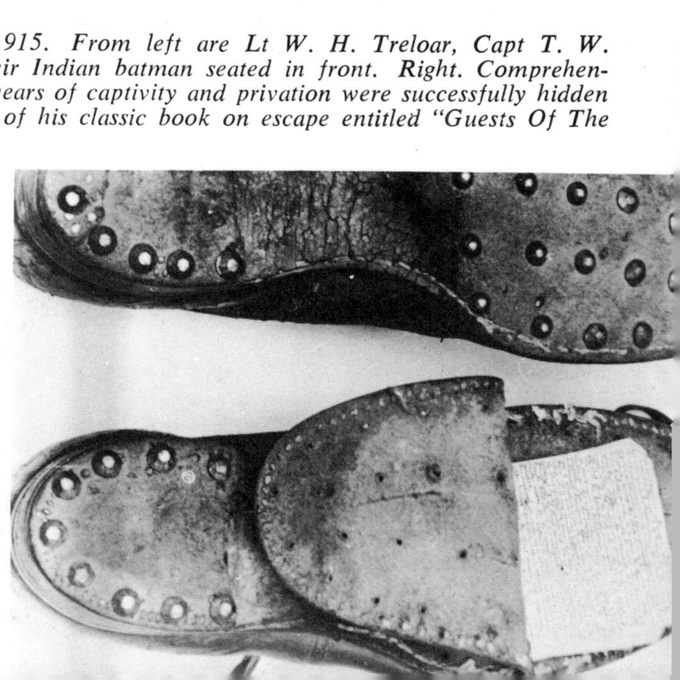

BE12a

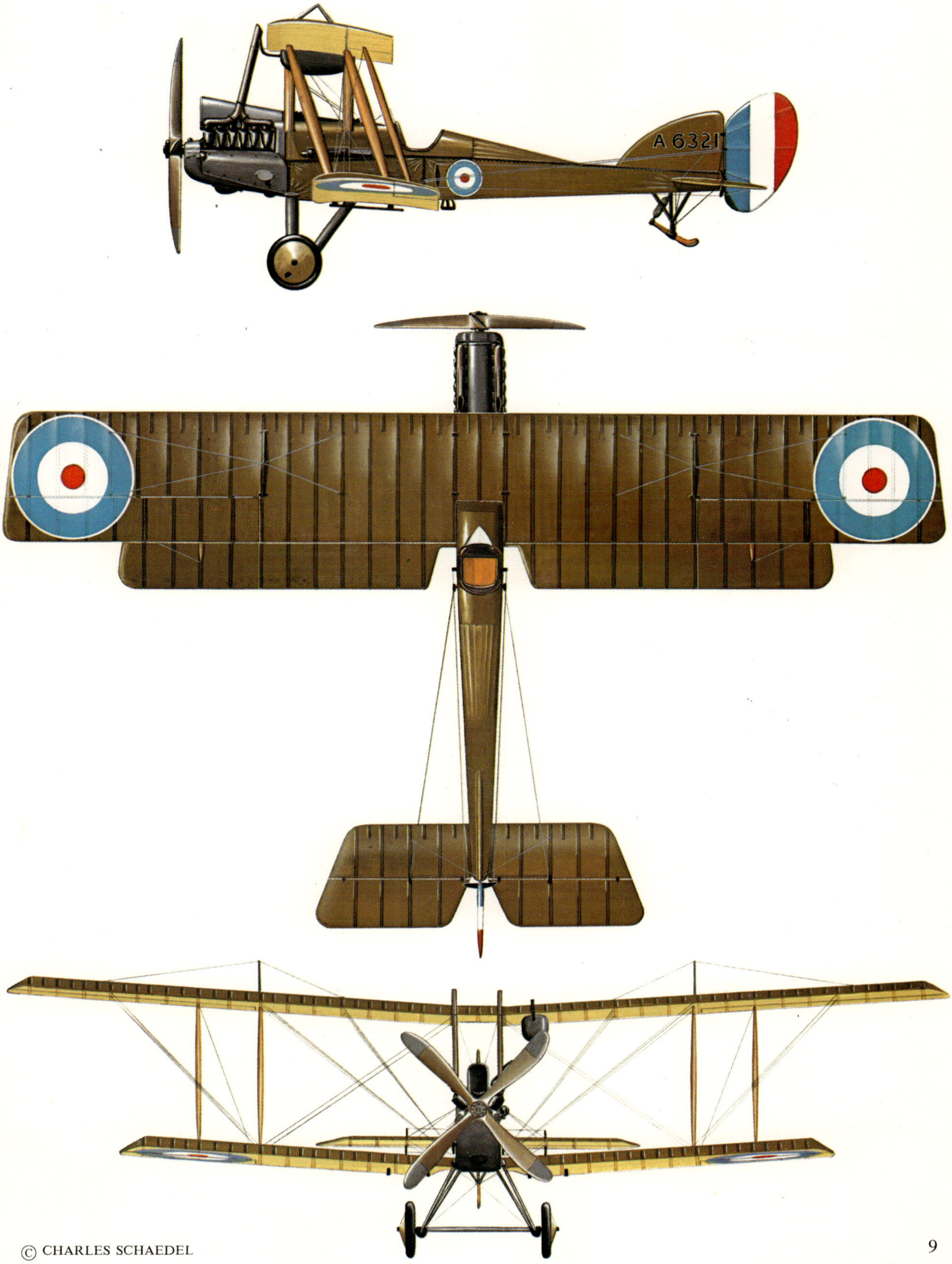

Among the large variety of aircraft used by No 1 Squadron AFC was BE12a, serial A6321. Relatively unsuccessful in France, the type was nevertheless put to good use in several roles in the desert theatre.

BRISTOL FIGHTER

Re-equipped with Bristol Fighters, No 1 Squadron soon established a definite superiority over the enemy. Among the pilots who flew B1146 was Lt Ross Smith, later Sir Ross Smith of England-Australia fame.

A pre-war designer and builder of his own aircraft, Air Mechanic A. R. Betteridge studies the features of the Point Cook Bleriot. Right. A BE2a at Point Cook displays its CFS serial number.

The French Caudron G III was better employed as a trainer at Point Cook than as an operational machine in Mesopotamia, where the shortcomings of a similar aircraft caused the loss of Lt Merz. Right. Splicing the rear fuselage longerons of one of the CFS Deperdussins.

A far cry from today's flight simulators, this "balancer" developed by Point Cook ground staff nevertheless provided a useful introduction to the "feel" of flying. Right. Two Curtiss Jenny aircraft as operated by "Billy" Stutt for AFC pilot training, photographed at Richmond, NSW, in the immediate post-war period. In the background are two Caudron trainers.

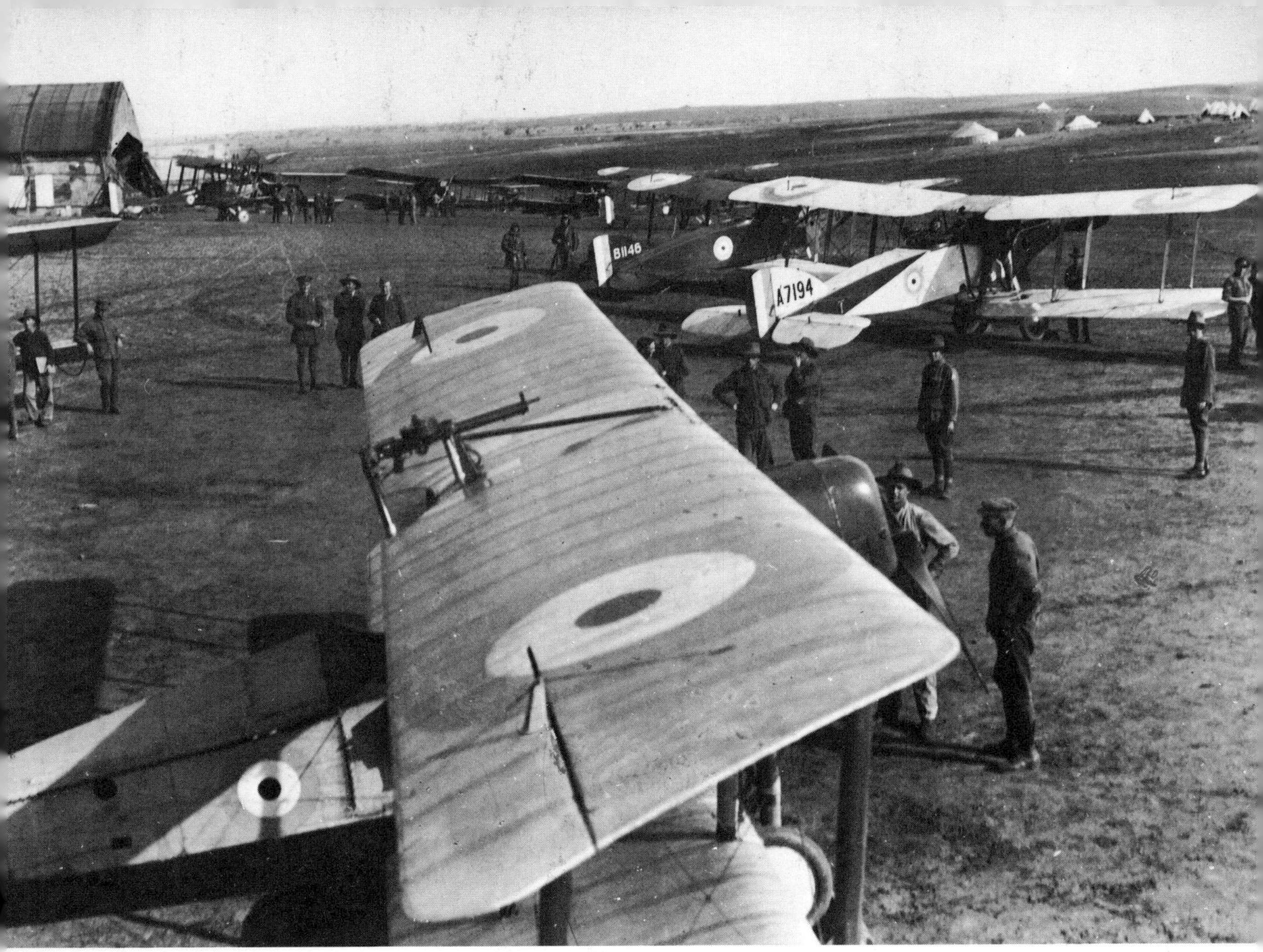

Desert Air Force! A variety of No 1 Squadron AFC aircraft assembled at Mejdel in Palestine in January 1918. Types represented are Martinsyde, Bristol Fighter and BE12a.

CHAPTER 2

THE DESERT WAR

As the campaign in Mesopotamia neared its tragic conclusion, the first complete Australian Flying Corps unit was being formed at Point Cook, following the suggestion of the British authorities that the Dominions might like to provide and administer their own service squadrons. Australia was the only country to act upon this suggestion, and No 1 Squadron, AFC, left Melbourne on 16th March 1916 for service in Egypt alongside the British units in that area. Their aircraft and associated operational equipment was to be supplied to them by the RFC, but it was six weeks after their arrival in mid-April before they received their first aeroplanes. The delay, however, was of little consequence, because their elementary training was of no value in the hard school of actual operational flying, and the majority of the pilots and observers went on to England to learn the finer points of air combat, photography and bombing, in which all-round proficiency was to be a real asset in their future work. The specialised division of fighting, bombing and reconnaissance units which was then being developed on the Western Front was not applied to the desert theatre until much later, and the pilots were called upon to perform a great variety of duties as soon as they began operations. Similarly, the Australian mechanics were attached to the RFC desert squadrons to gain the experience necessary to keep their aircraft flying in conditions far removed from anything they had previously encountered. At this point they began to appreciate that their formidable task was only slightly better in terms of equipment and supply than that of the Mesopotamian force, and that initiative and skill in improvisation were necessary qualities of high importance.

During this period the squadron command changed hands continually. Lt-Col E. H. Reynolds had brought it from Australia and handed over to Maj H. D. K. Macartney, who in turn was replaced by Maj A. A. J. Broun. Finally Maj T. F. Rutledge took over from Broun in time to become the first commander of a full Australian Flying Corps operational squadron. Its first assignment was to have been the replacement of one of the RFC squadrons, but it split up once again, its separate flights taking over widely dispersed areas. "B" Flight under the experienced Capt Oswald Watt moved out first, followed by "A" Flight under Capt W. Sheldon and finally "C" Flight under Capt R. Williams, the first pilot to qualify for his wings at Point Cook. These moves began about mid-June in 1916, and were followed by a succession of others until the rapid advances of the land forces towards Palestine finally resulted in the squadron being re-united towards the end of the year.

Numerically speaking, the aircraft position had improved since the time when some of the Austra-

A World War I scramble! An approaching enemy aircraft causes a hurried take-off by BE2c aircraft of No 1 Squadron from Kantara aerodrome in Egypt.

lian pilots had to fly RFC aircraft to gain experience, but in quality their machines left much to be desired. The main type in use was the BE2c, a reconnaissance aircraft which had been designed with built-in stability, making it pleasant to fly but a deathtrap in circumstances which required rapid evasive action. Coupled to this, it had no synchronising gear to enable a fixed machine-gun to be fired forward through the revolving propeller blades, and relied for its defence on a movable Lewis gun operated by the observer. This worthy gentleman occupied the forward seat completely surrounded by wires and struts, so it can be readily appreciated why the type was mainly responsible for the introduction of the term "Fokker fodder" on the Western Front. Although the German opposition in Egypt was not quite as sophisticated as that encountered in France, at least they had the advantage of well-armed scout aircraft and two-seaters in which the observer could take an active and offensive part in any combat. The practice on both sides was to equip the squadrons for bombing and reconnaissance duties and allot a few single-seat machines for escort purposes, the latter being flown by pilots within the units who showed a special aptitude for this type of work. This system worked quite well, but the Australians' aggression was not confined to pilots of the single-seat machines as witnessed by various attempts to devise effective gun mountings for the BE's. Both officers and mechanics were busy in this respect, and mountings were devised to allow operation of the Lewis guns by the pilot, who was often required to fly alone because the weight of bombs precluded the carrying of an observer. Attention was also given to devising an effective synchronisation gear.

Apart from a few DH1 and DH2 pusher scouts which allowed a clear field of fire ahead of the machine, the other types such as the Bristol Scout, the Martinsyde and the Vickers single-seaters relied on various forms of fuselage and wing mountings for their guns, none of which was entirely satisfactory. Germany had shown the way with the first use of a really efficient gun system, and had reaped the benefit while the Allies strove desperately to develop synchronising gears of their own. (One such system coming into use at this time was the Sopwith-Kauper gear developed by the Australian Harry Kauper while employed at the Sopwith works in England.) It is doubtful whether the RFC and AFC units in Egypt would have been officially advised of the latest technical developments, and even less likely that the authorities would have placed them very high on the priority list for allotment of such equipment at this time. The aircraft which provided No 1 Squadron's offensive element were initially Bristol Scouts, which came on strength toward the end of 1916, to be followed by Martinsydes early in the next year when the performance of the Bristols was found wanting. The Martinsyde was a relatively large aircraft originally designed as a long-range reconnaissance machine, with a single Lewis gun mounted on the top wing. This type of machine served with the squadron throughout 1917 and into 1918, doing a good job despite the unfortunate tendency

Lt A. T. Cole of No 1 Squadron AFC preparing a BE2c for flight with the assistance of ground crew members G. Turner and J. Luff. "King" Cole later distinguished himself further with No 2 Squadron in France and the RAAF. Right. A full bombload slung beneath a BE2c and ready for presentation to the enemy.

A Martinsyde aircraft of No 1 Squadron AFC. The men behind the fuselage emphasise the relatively large dimensions of the type.

of its Beardmore engine to blow back through the carburettor on opening the throttle after idling, with a consequent risk of fire.

These early types with their crude armament provided the necessary incentive for the inventors within the squadron, among them Lt L. J. Wackett, who later played a leading part in the establishment of Australia's own aircraft industry, but at that time a pilot with No 1. He used one of the Bristols for experiments with a rod and cam interrupter gear, similar to the system used by the Germans to avoid the embarrassment of shooting holes in the propeller. However, Wackett's scheme was not entirely successful. A technically sound yet simple hydraulic synchronising system was devised by Allan Betteridge, a squadron mechanic, but his submissions to the authorities produced no enthusiastic response, although a hydraulic system developed by a Rumanian was adopted almost universally at about this time.

Apart from the single-seat aircraft, the BE reconnaissance types comprised the main equipment of the squadron, with the BE2c eventually being replaced by the BE2e, an improvement of no great consequence. A later addition to the strength was the BE12a, a single-seat version of the 2e. This had been introduced into service in France in the vain hope that its performance would be sufficient to match that of the German fighters and reverse the heavy losses being experienced by the British reconnaissance units. Having failed miserably to achieve the hoped-for success, the BE12a was relegated to operations in so-called less important areas and some of them found their way into No 1 Squadron. Nevertheless, all these types were employed with a great deal of success by the Australian unit throughout 1917, until the rapid escalation of the war in this area resulted in the arrival of some RE8's. Finally, in January 1918, the squadron was re-equipped with Bristol Fighters taken over from one of the neighbouring RFC squadrons. At last the Australians found themselves with an aircraft that had the ideal all-round capabilities necessary for their type of operation, and they wasted little time in exploiting them to the full.

Before this happy state of affairs, however, No 1 Squadron ranged far and wide over the desert areas around the Suez Canal, initially to the west and then to the east as the German-Turkish forces advanced toward the canal before being stopped and driven back by the Light Horse and infantry units. The tasks of the airmen were reconnaissance, bombing and photography, aerial clashes with the enemy being relatively few. This was due partly to the inability of the British aircraft to catch the superior German machines and partly to the reluctance of the enemy at that time to engage in this type of warfare. But it was no easy war! Patrols were long, with the ever-present danger of a forced-landing as a result of engine failure or the heavy and uncomfortably accurate ground fire which was often encountered. Although a successful landing was often possible under these conditions, the unlucky airmen then had to contend with the hostile Arab tribes and the Turkish soldiers, together with

A Bristol Scout used by Lt L. J. Wackett in experiments with machine-gun synchronisation. A metal disc replaced the engine and propeller for testing purposes. Right. Even an experienced pilot like Capt Oswald Watt was subject to the vagaries of the desert winds and the Martinsyde.

Lt C. R. Andrews in a BE2c of No 1 Squadron AFC. The gadget on the side of the cockpit is a bombsight, NOT a handbrake as seen on contemporary motor cars and horse-drawn vehicles!

miles of some of the most inhospitable country in the world before they could reach their own lines. Little wonder that daring and dangerous rescues were attempted wherever possible under circumstances such as these, and a spate of these incidents in March 1917 culminated in Lt Frank McNamara winning the only Victoria Cross awarded to an Australian airman in World War I.

Early in the month Lt J. V. Tunbridge, of No 1 Squadron, was forced to land his damaged aircraft in the desert, and then began a long walk home through an area peopled by Arab tribesmen who were more inclined to slit the throat of an infidel than shake his hand. Fortunately he was located from the air by Lt P. W. Snell, an Australian serving with the RFC and temporarily attached to No 1, who made a successful pick-up of the exhausted Tunbridge. Considering that it was usually necessary to leave the observer behind when carrying a full bomb load of something like 240 lb, it was no small achievement for Snell to lift his BE into the hot desert air from unprepared ground with the combined weights of his own observer Morgan and Tunbridge exceeding that figure by a generous margin. The same pilot attempted to repeat the performance a couple of days later when Lt L. W. Heathcote was forced down, but he was unable to find a suitable landing place and Heathcote was taken prisoner. On 19th March an RFC aircraft was brought down by ground fire and Lt R. F. Baillieu carried out a successful rescue, his observer at the time being Lt Ross Smith, later to graduate as a pilot and become one of the leading lights in the squadron.

The next day, following an attack on enemy rail communications, Capt D. W. Rutherford's

By comparison, this ingenious experimental bombsight made by Air Mechanic Betteridge compensated for variations in aircraft attitude during aiming. It was not adopted by the authorities. Right. An informal discussion between pilot and mechanic concerning the performance of the Bristol Scout (or possibly the removal of a speck of the ever-present sand from the pilot's eye).

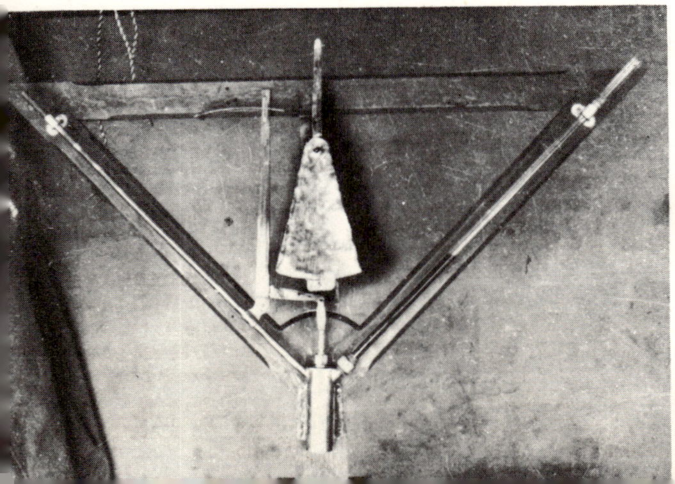

A No 1 Squadron Martinsyde taxis in after a patrol. This was a danger period, for a sudden opening of the throttle after prolonged idling could start a fire caused by carburettor blowback.

BE2c was brought down near a force of enemy soldiers and he was in danger of imminent capture. Lt McNamara, although badly wounded by a bullet from the ground which had come up through the bottom of his seat, saw the situation and immediately landed his Martinsyde to attempt a pick-up. With Rutherford hanging on grimly, he began the take-off, but his wound prevented him from exercising complete control under these abnormal conditions and the Martinsyde crashed. Fortunately, the two men suffered no further injury, and after setting fire to the wreck they managed to start the engine of Rutherford's BE. In the face of the rapidly approaching Turks, McNamara was this time successful in getting off the ground and flying the 70 miles back to base. It was this action which earned him Britain's highest award for valour.

The German airmen now began to adopt a more belligerent attitude to the British machines in the air, although they had always been active in attacking ground targets. Most of the Australian losses had been brought about by ground fire, but 28th March saw No 1 suffer its first fatality in aerial combat. While engaged on artillery observation, a BE flown by the unlucky Capt Rutherford was attacked by a German Aviatik two-seater, and he and his observer, Lt W. R. Hyam, both suffered wounds before the enemy machine was driven off by another Australian aircraft. Rutherford succeeded in reaching home, but unfortunately his observer succumbed to his wounds a couple of days later. Hyam had been one of the volunteers from the Light Horse and Anzac divisions with which the Australian squadron maintained its establishment throughout its service in Egypt and Palestine. (Rutherford continued to fly with No 1 until the first day of May in the following year when he was again brought down by ground fire. A pick-up attempt by Lt F. W. Haig ended in a crash which resulted in the two pilots and their observers finishing the war as prisoners).

It became apparent that the action which resulted in the loss of the Australian observer was the beginning of a new approach by the German airmen. Reinforced with more and better aircraft, they began to show a determined resistance to the patrols of the AFC pilots, but despite their unsuitable equipment the Australians never refused a fight, although the results were usually indecisive. The main danger was still accurate ground fire which caused the squadron several losses during the next couple of months, together with a near miss when Capt Williams essayed a successful pick-up of Lt A. T. Cole on 21st April. Cole himself took part in an extreme example of these rescue operations during June, when eight machines were returning from a raid on Jerusalem. One BE was forced down with engine trouble, and Cole in his Martinsyde and another pilot in a BE landed to render assistance. After burning the disabled aircraft, the pilot was taken off in the BE, but this also came down before reaching base. Cole's attempt to rescue the two stranded men failed when he wrecked his undercarriage, and the three began a perilous walk home. Fortunately, they were eventually rescued by Light Horse units. Of the remainder

Desert rescuers! Capt F. H. McNamara, VC (on right) and Lt R. F. Baillieu in hospital in Egypt subsequent to their pick-up exploits on succeeding days as described in the text. Right. BE2c aircraft of No 1 Squadron AFC being prepared for a raid at Kantara in Egypt.

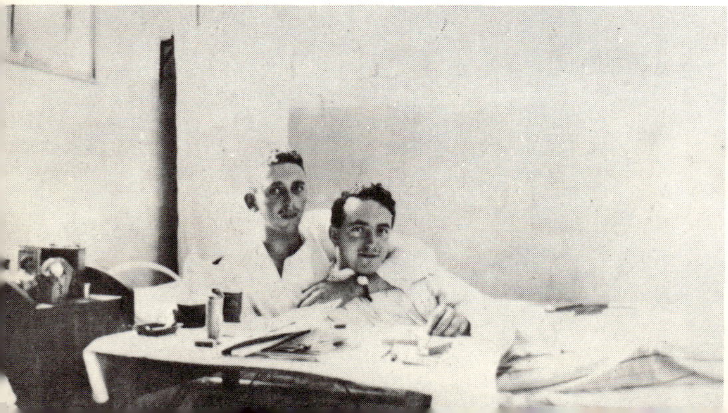

Proof that the desert was not always dry is evident in this photograph of a BE2e taxying through water-filled ruts after a heavy rainstorm.

of the formation, two more were forced down by engine failure but the pilots were rescued by their companions and returned safely to base. These engine failures resulted not from enemy action but from the extreme operating conditions under which the desert units worked; conditions which made it very difficult to maintain the squadron at strength.

No 1 had already suffered a serious blow on the previous day when Lt J. S. Brasell was shot down in the squadron's first BE12a, the type that was hopefully expected to give them a fighting chance against the enemy. Further supplies of BE12a's failed to provide the hoped-for superiority, but with the Martinsydes they provided a determined if somewhat ineffectual deterrent to the enemy machines. However, during the latter half of 1917 the winds of change began to blow favourably for the British and Australian airmen. Bristol Fighters began to make their appearance in the hands of a recently-arrived RFC squadron, and No 1 exchanged its previous multitudinous duties to become a tactical bombing unit after the arrival of a further RFC squadron. The strength of the Australian unit was also increased from within its own ranks and the Light Horse divisions in order to provide for shortages incurred by casualties and to replace personnel transferred to the new AFC squadrons. Among the first to leave were some of the attached Australian members of the RFC, such as Maj Rutledge and Capt Watt, with the result that in May 1917 Capt Williams took over the command of No 1, being the first member of the AFC to command a full operational squadron of that service.

The build-up of Allied air strength during this period resulted in a comparable reinforcement of the opposing German forces in October, and the stage was set for intensive aerial activity during the offensives planned for the end of the month against Beersheba and Gaza. Shortly before this, No 1 Squadron began to receive the long-awaited RE8, together with an improved Martinsyde with a more powerful engine. These and the BE12a's were used to good effect during the opening days of the offensive. Although air opposition was very light, considerable successes were achieved against their immediate foe by bombing attacks against several aerodromes. This resulted in the destruction of a large number of enemy aircraft on the ground. This pattern of fighting occupied the squadron for the remainder of the year, but 1918 brought further change and their most rewarding efforts since becoming operational.

To meet the ever-increasing threat of the German fighters, the RFC Bristol Fighter squadron was re-equipped with the single-seat SE5a, and the Bristols were taken over by the Australians. The RE8, excellent for the close-support work necessary at the time of its arrival in the desert, was found wanting when it came to the long-range reconnaissance patrols in preparation for the coming advances deep into Palestine, but the two-seat Bristol was ideal because of its range and offensive ability. Mixed formations were still in use until the end of March,

Rescues were not confined to personnel alone. A BE2e is escorted back to base after being fitted with a new undercarriage and propeller out in the desert. Right. These Martinsydes were among a number of No 1 Squadron aircraft badly damaged by one of three waterspouts from the sea which left a trail of destruction at the end of 1917.

Under very trying weather conditions, ground crew members prepare to start a veteran RE8. What was said to an uncooperative engine on these occasions can best be left to the imagination.

but even in a BE there was a greater feeling of security in the knowledge that the escort Bristol was well able to take care of any German fighter that decided to gate-crash the party. This form of security made it possible in mid-January for five aircraft of the squadron to photograph an area of over 600 square miles behind the Turkish lines in a period of a fortnight under the protection of two Australian Bristols and one from the RFC squadron. However, as an instance that all was not lost if the enemy succeeding in eluding the escort, Lt Len Taplin drove off a troublesome Albatros which attacked his BE12a at a most inconvenient moment, when he had parts of his aerial camera spread all over the cockpit to carry out some repairs. After a brisk engagement which resulted in the German leaving the scene in a vertical dive, Taplin calmly re-assembled the camera and continued his work.

The advent of the Bristol Fighter also brought with it the first confirmed victory to be recorded by No 1 Squadron. This was gained on 3rd January 1918 when an Albatros scout was seen to crash after being attacked by a Bristol flown by Lt R. A. Austin with Lt L. W. Sutherland as his observer. Aerial supremacy was rapidly established until finally there was no effective enemy reconnaissance available to warn of the imminent large offensive being prepared for mid-September. During the preceding months the Australian Bristols were utilised for every type of operation possible and wreaked havoc with bombing, ground strafing and aerial combat, all combined with their initial duty of long-range reconnaissance. The number of enemy aircraft destroyed was not great by Western Front standards, but neither were the numbers of aircraft employed, and the total of 15 destroyed in the two months before the opening of the final offensive was no mean tally. In addition, many more were listed as "driven down", which was the term employed when the enemy pilots were forced to break off the combat and head for the nearest patch of suitable landing ground, whether it be an aerodrome or a spot in the desert. The fact that these victories by No 1 comprised the full total on the whole front is explained largely by the ability of the Bristols to range far and wide over the enemy back areas to catch his aircraft on their home ground. Eventually the Germans were forced to adopt high-altitude

A slightly bent BE12a being inspected by ground crew prior to salvage. The desert sands at this point were obviously hot but certainly not soft. Right. This stone bears the following inscription beneath the names of AFC and RFC airmen brought down behind the enemy lines in Palestine:— "This monument generously erected to one of them by their enemies was discovered and restored by their friends, January 1918."

A BE12a, serialled A6321, leaves a small plume of dust behind its tailskid as it taxis across the seemingly limitless expanse of its aerodrome.

forays in an endeavour to gain information, a most ineffective method. During this period of intense activity the Australian losses were surprisingly light, although heavy and accurate ground fire still remained one of the greatest hazards. The small number of casualties suffered in the aerial duels were mainly caused by the German two-seaters, despite the introduction of an enemy fighter squadron in an attempt to restrict the damaging work of the British aircraft.

Apart from the regular duties allotted to No 1 Squadron, members of the unit also took part in other activities which played a significant role in the outcome of the campaign. The first of these was to lend support, both practical and morale-building, for Col Lawrence and his Arab forces, and this support was given by various Australian crews from mid-1918 until the end of the desert hostilities. In addition, Australian mechanics were supplied to help maintain two BE12a aircraft which were permanently attached to this desert force along with several British officers.

The other outside activity centred around the only Handley-Page bomber to appear in the Middle East during the war. When this large twin-engined machine arrived from England at the end of August 1918 it was allotted to No 1 Squadron, where it became the mount of Ross Smith, by this time an air fighter of no mean skill. Apart from the propaganda value of the large bomber in convincing the Arabs that it would be in their best interests to side with the British forces, its most important function was to be used during an advance operation to signal the opening of Gen Allenby's final offensive in Palestine. Accordingly, Ross Smith set out in the early hours of 19th September for the enemy communication centre at El Afule, accompanied by Lts Mulford, Lees and McCann. Sixteen 112 lb bombs effectively silenced the telephone exchange and badly damaged the railway, thereby leaving the inland Turkish forces in complete ignorance of the attacks against their coastal army. These attacks were pressed home with such determination that large numbers of the enemy were soon retreating in rapid disorder.

The retreat was hastened and further disorganised by relays of Bristols, which maintained a steady attack with bombs and machine-guns in the face of heavy defensive fire. The speed with which the ground forces advanced can be gauged by the fact that the crews of two crashed aircraft were only in captivity for a few hours before being released by the cavalry. Enemy air opposition was virtually nil, with the British fighters successfully restricting any attempts by the German airmen to interfere with the work of the ground-attack aircraft. As the organised anti-aircraft fire was steadily reduced, strafing planes were going about their work at heights as low as 150 feet.

By the third day of the attack, continued bombing and strafing had completely disorganised the fleeing Turks, who were still cut off from their blissfully

"To remove engine from airframe in shortest possible time, drop aircraft onto aerodrome from height determined by experiment." The result of a heavy landing by RE8, serial B5853. Right. Prior to its mishap, B5853 is shown in its initial splendour on joining No 1 Squadron.

Double deadliness! Twin Lewis guns of a Bristol Fighter in the hands of Cpl N. P. B. Jensen, a former Danish sailor. He and his pilot, Lt J. K. Curwen- Walker, were unfortunately killed later in a take-off accident. Right. A Bristol Fighter of No 1 Squadron undergoing maintenance in the desert.

ignorant comrades to the east. This third day also witnessed one of the most terrible examples of the effective use of air power ever seen up to that time when an Australian air patrol discovered the major portion of the enemy transport seeking to escape along the road from Balata which ran through the Wady Fara toward the River Jordan. By bombing both ends of the packed column, the Bristols bottled up the enemy force along a nine-mile stretch of road preventing escape to either side and then, together with the British squadrons, proceeded to annihilate it during that and the following day.

For the Turks and the Germans this was the beginning of the end. The headlong advance of the cavalry and armoured units kept them on the run, continually under attack from the air, until they had no alternative but to call for an armistice on 31st October. During this final retreat No 1 Squadron continued its deadly work against the ground forces and railways, although some air opposition was encountered as they over-ran the last German aerodromes. One of the last actions of this type occurred on 23rd October when Lts Harper and Lilly attacked six aircraft on Muslimie aerodrome, which was evacuated by the Germans later that day.

After the Armistice, No 1 Squadron remained in Egypt until March 1919 when it returned to Australia, leaving behind a proud record of a job well done under extreme circumstances, and having set an example which was to be followed just over 20 years later by the RAAF squadrons in the Middle East during World War II.

Thoroughbred lines! The Bristol Fighter earned the same praise that ex-Light Horsemen with No 1 Squadron had formerly bestowed on their equine mounts.

Bristol Fighters being serviced in the open at Ramleh in Palestine in 1918.

A Rolls-Royce Falcon engine from a Bristol Fighter being run-up on a test stand in Palestine. Right. Armourer F. W. Haig with various weapons in Egypt. Haig later graduated as a pilot with the squadron, eventually being taken prisoner in 1918.

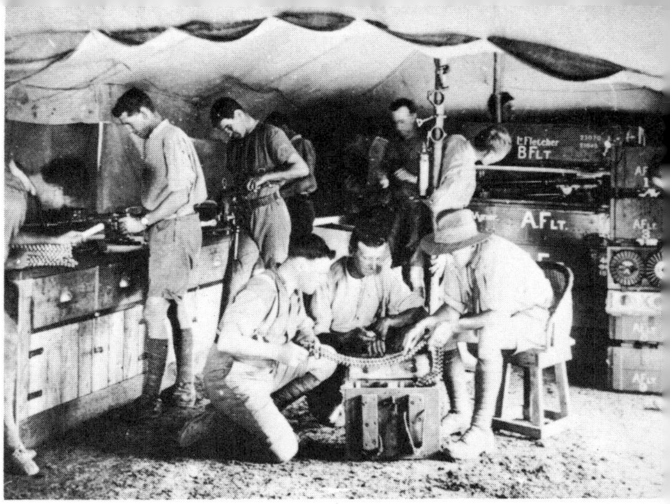

The radiator and sheet-metal repair section of No 1 Squadron. Right. The armament section at work.

The Flying Corps equivalent of leaning on a shovel? What better support for a tired back than the broad propeller blade of a Bristol Fighter! Right. Wings folded and looking deceptively docile, the only Handley-Page 0/400 used in the desert theatre is shown being manhandled by ground staff of No 1 Squadron.

A salvaged Martinsyde after being brought back to base on a trailer. Right. Preparing to test a Beardmore engine from a Martinsyde. Sgt Kilburn, at right, was in charge of the operation.

An RE8 and crew of No 3 Squadron AFC. Ungainly and unstable, these aircraft in the hands of experienced crews nevertheless achieved an outstanding record for the first AFC unit to operate in France.

CHAPTER 3.

THE EYES OF THE ARMY

No 1 Squadron lost a number of its experienced members when it assisted in putting more AFC squadrons on a war footing for operations in France. Two of these units came from Australia, the third from Egypt. In order of entering operations they were: No 3 Squadron, AFC (originally 69 Squadron, RFC); No 2 Squadron, AFC (68 Squadron, RFC); and No. 4 Squadron, AFC (71 Squadron, RFC). These designations followed the system which originated with No 1 being listed as 67 Squadron, RFC. First to arrive from Australia during the last week of 1916 was the unit which saw action as No. 3, under the command of Maj D. V. J. Blake. The air crews immediately embarked upon a training schedule to make them familiar with the unspectacular but highly important work of reconnaissance and artillery observation which by now had developed into a specialised field.

As a result of this the training occupied as many months as the original No 1 men had been allowed weeks but, finally, in August 1917, the RE8 aircraft and men of the new squadron set out on the first leg of the journey to France, only to suffer casualties before leaving the English shore. Lt F. C. Shapira and A/M W. D. Sloane were killed when their machine crashed on take-off after a forced landing in Kent. Bad weather caused further delays until 9th September 1917, when the pilots at last pointed the long noses of their aircraft out across the English Channel and set a course for the war zone. After an overnight stop at the distribution centre of St Omer they proceeded to their designated base at Savy the following day. From here they worked in conjunction with two RFC squadrons, feeling their way around their new environment until they had sufficient experience to be allotted a sector of their own. This move came, early in November, to Bailleul where they found themselves ranging over the Messines area in support of the Australian infantry.

Here the crews of the ungainly observation machines found themselves prime targets for enemy fighter formations determined to make it as difficult as possible for the Australian airmen. The men of No 3 Squadron were convinced that they had at least an equal right to that particular patch of sky and they fought with every ounce of skill at their command to defend that right. Many a numerically superior German force was beaten off after attacking a solitary RE8, finding the elusive victim ready to give a good account of himself until the arrival of help. The offensive spirit of the squadron is indicated by the fact that during its 12 months of serious operations more than 50 enemy aircraft were forced out of the sky, 16 of them being totally destroyed in the air and the rest either shot down out of control or forced down. The first victory to the squadron was recorded on 6th December when the pilot of a DFW two-seater decided to interrupt the artillery observation work of "A" Flight commander Capt W. H. Anderson. Before the German had time to fire a shot the RE8 observer, Lt J. R. Bell, emptied a drum of Lewis gun ammunition into him and the DFW crashed into its own lines.

Less than a fortnight later Australia gained a war trophy which is still an exhibit of great interest in the Australian War Memorial in Canberra. Unfortunately the action also cost the lives of two members of No 3 Squadron. The trophy, a German Albatros DVa fighter, was one of a number which attacked the RE8 of Lt J. L. Sandy and Sgt H. F. Hughes on 17th December. The Australians fought so relentlessly that this aircraft was forced down in British territory and was salvaged by personnel of No 3, eventually to be handed over to Australia as a prize by the British authorities. While the badly wounded German pilot desperately sought refuge on the ground, however, Sandy and Hughes continued to hold off the remaining attackers until the arrival of another RE8 reduced the odds to a more acceptable level. The loss of their comrade evidently made the Germans eager for revenge and the fight continued for another 10 minutes, an unusually long period for an action of this kind and one that must

Prior to AFC operations in France, selected officers served in RFC units for experience. Lt N. L. Petschler, seen here with an RE8 of 42 Squadron RFC, later flew with No 3 Squadron AFC and eventually became a flight commander with No 4 Squadron AFC.

have seemed like hours to the sweating crews of the heavy observation machines. At this juncture the arrival of a third Australian aircraft persuaded the German formation to break off the attack and they left the area to the three RE8's. The RE8's reformed completely unaware that the aircraft of Sandy and Hughes now carried a dead pilot and a mortally wounded observer. Flying in large sweeps as though endeavouring to complete the mission of its crew, the faithful old RE8 finally ran out of fuel and gently brought the two bodies to ground in a relatively soft crash-landing 50 miles away. The news of their fate reached the squadron the following day.

During this period foul weather greatly restricted the activities of the air forces, but at every opportunity aircraft of No 3 took the air to photograph, bomb and observe the enemy. And, when occasion demanded, they engaged in the more personal duels of aerial combat. As the winter weather began to moderate into the spring of 1918 it became evident that the Germans were preparing for a massive offensive, and the corps squadrons were kept busy recording with their cameras every aspect of the enemy preparations. They were also kept busy avoiding the attentions of the German aircraft whose main aim was to ensure that these important plates were never exposed. Fighter escorts were provided for the protection of the two-seaters but, with the increase of aerial activity on both sides, it was impossible for the escorts to be everywhere at once and often the observation crews were forced to delay their work to defend themselves.

Here at least they could see their foe and act accordingly; unlike the invisible anti-aircraft shell which suddenly coughed and spat its deadly fragments of red-hot steel at them. Normal operations usually allowed the pilots to take evasive measures through changes in course and height, dodging their way between the stinking black bursts of high-explosive; but photographic missions required straight and level flying which was akin to running the gauntlet. Coupled to this was the ever-present danger of being struck in mid-air by the artillery shells, both friendly and enemy, which sometimes traversed the same region of air. These were normally unseen and unheard, although there were instances of airborne howitzer shells being seen momentarily as they turned over at the top of their high-angle trajectory. The Australian unit lost two machines to shell-fire early in the year but more than made up for them in the amount of damage they caused. The loss of Lts C. Donahay and J. R. Blair as a result of anti-aircraft fire on 26th January had already been paid for on the previous day with the destruction of a German two-seater by Lts C. C. Matheson and C. T. Brown.

As if the constant tension of operating under such trying conditions was not enough, No 3 was also subjected to attack on the ground as the enemy began to prepare the way for the March attack. Bailleul became a target for the German artillery and heavy bombers, with the aerodrome a prime target, and toward the end of March a move had to be made to Abeele. Here, another victory was recorded when Lts R. G. D. Francis and R. Hainsworth emerged the victors after a long fight against

The German Albatros DVa shot down by Lt J. L. M. Sandy and Sgt H. F. Hughes of No 3 Squadron. This machine is now the property of the Australian War Memorial at Canberra.

An Australian RE8 displaying its identification letter on the fuselage and upper wing centre-section. A new starboard upper aileron had recently been fitted as is shown by the incomplete cockade.

a pugnacious DFW two-seater which they forced across the British lines and finally destroyed near their own aerodrome.

Early in April a further move to Poulainville took the squadron south to cover the Somme sector and within a fortnight it had become involved in one of the most controversial incidents of World War I. Late in the morning of 21st April, Lts S. G. Garrett and A. V. Barrow, together with Lts T. L. Simpson and E. C. Banks, were attacked near Hamel by a number of enemy Fokker triplanes. The two RE8 crews fought them off until the imminent arrival of a British fighter formation forced the enemy to break away. The Australians resumed their reconnaissance, unaware that they were among the last people to see Germany's almost legendary airman, Baron Manfred von Richthofen, alive in the air. Shortly afterward the German ace lay dead in the wreckage of his maroon triplane near the Australian infantry in that area. Whether he was killed by a bullet from the guns of one of the British fighters or by ground fire from the Australian positions is an argument which is still being fought almost as hotly as his final combat. His body and the remains of his aircraft were taken into No 3 Squadron's base from where he was given a military funeral consistent with a gallant foe of his calibre, complete with guard of honour and wreaths from various units of his former enemies. His aircraft, far from being completely wrecked in the crash, suffered most at the hands of the souvenir hunters, and while some of the less transportable parts were kept for display purposes, there must be many homes in Australia which have a small piece of its fabric tucked away among dusty and unrecognised relics.

This period in the Somme area brought an increase in the amount of bombing and attack missions carried out by the squadron, a job rendered doubly dangerous by the fact that concentration had to be equally divided between the ground and the surrounding sky. Here, the two-seater had an obvious advantage; the presence of a watchful observer allowed the pilot to apply himself fully to the job in hand until warned by a thump on the back from his companion that enemy aircraft were about to break up the party. These attacks were frequent, the cloudy weather enabling the enemy to stalk the solitary observation machines before launching a sudden sortie from cover.

Despite the fact that an RE8 often had to contend with something like half a dozen fast enemy fighters, losses from these combats were relatively light and in quite a few cases resulted in another victory for the squadron score. One of the flight commanders, Capt E. J. Jones, seemed to attract German fighters in the same way that a honeypot attracts bees, and he was instrumental in two of his observers adding

Officers of No 3 Squadron AFC.

RE8 aircraft of No 3 Squadron being serviced at Premont in France in October 1918. The white disc originally carried as a squadron marking was dispensed with in March of that year.

to the list with a Pfalz falling to the Lewis gun of Lt A. L. D. Taylor on 21st April (the day of Richthofen's death), and a Fokker triplane to Lt R. Hainsworth on 16th May. Unfortunately, another fight with triplanes four days later resulted in Jones being wounded and Taylor being killed, although the stricken pilot managed to reach his aerodrome. This injury to Jones completed a bad run for the flight commanders of No 3 during May, two having been previously lost to the squadron with the death of Capt Ralfe and the wounding of Capt Duigan on the 6th and 9th respectively (Duigan, as already mentioned, had been credited with the first flight in Australia in a locally-built aircraft, which is still preserved in Australia. Another lesser-known relic with which he was associated was a large German railway gun which he first spotted from the air and which, after its capture several months later, was presented to Australia as a war trophy).

In addition to the usual difficulties facing the observation crews, heavy fogs were often encountered. Caught in one of these, with no blind-flying aids to help him maintain the correct attitude of his aircraft, Lt W. V. Herbert found himself forced down to ground level on the wrong side of the lines. In imminent danger of being shot down or colliding with one of the high objects that continually flashed into view ahead of the madly weaving nose of the

The price of war. A BE12 propeller serves as a headstone for Lt O. G. Lewis, killed with Lt G. W. Best in the crash of their RE8 on 12th April 1918. Right. A unique and horrifying picture of an RE8 plunging earthward with its upper wing extensions folded back.

RE8

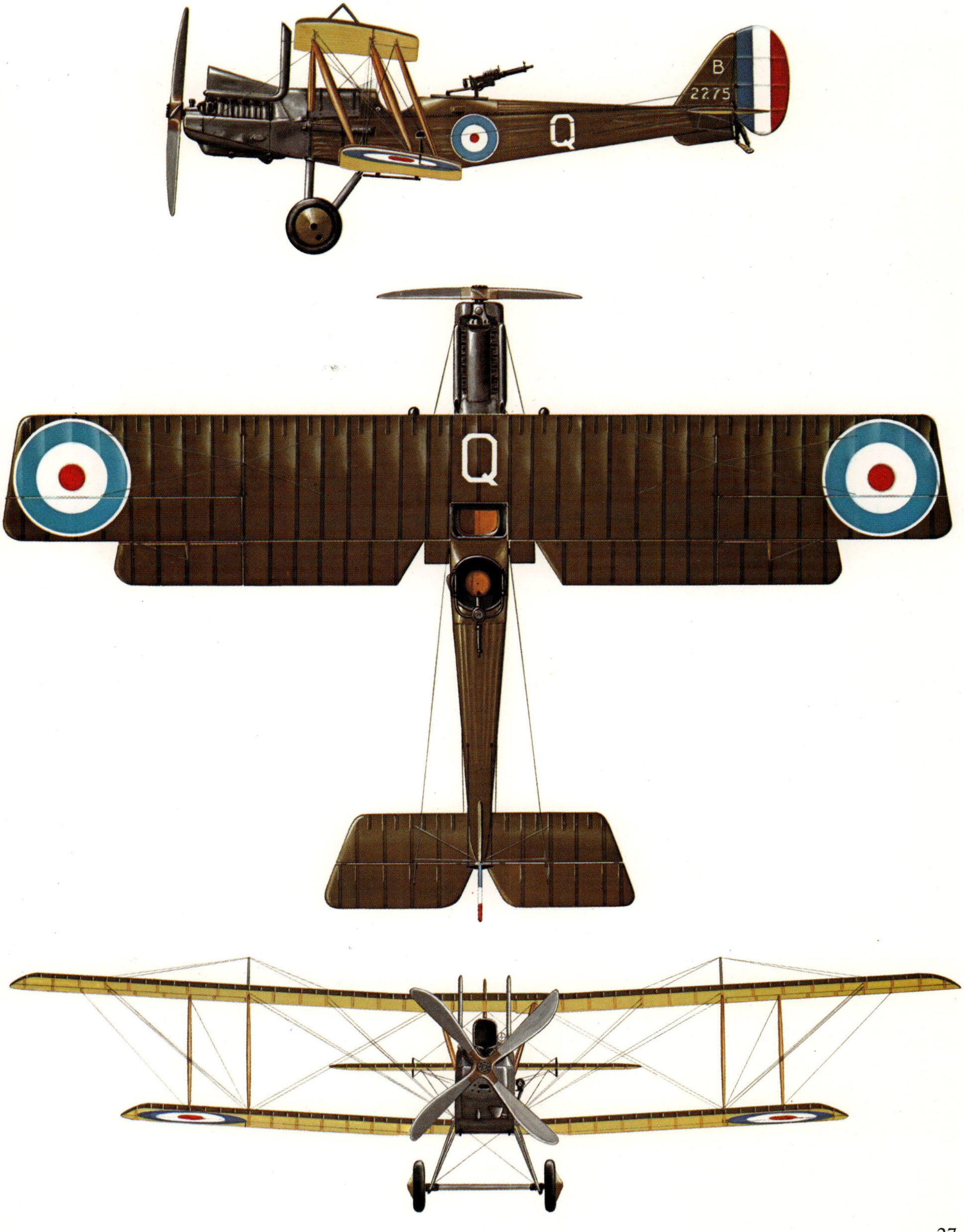

RE8, serial B2275, was typical of the machines which enabled No 3 Squadron AFC to become one of the leading Corps squadrons in France. This aircraft was flown by Lts R. C. Armstrong and F. J. Mart.

DH5

No 2 Squadron AFC began operations equipped with the DH5 scout, which it used to good effect on ground-attack work during the Battle of Cambrai. A9449 was the mount of Capt. G. C. Wilson, the leader of "B" Flight.

SE5a

In the closing days of 1917, No 2 Squadron converted to the SE5a, and used the boomerang marking illustrated until the end of March 1918. C9539 was flown by Capt H. G. Forrest, leader of "C" Flight, and this was the machine in which he scored several victories during the March enemy offensive.

SOPWITH CAMEL

No 4 Squadron also used the distinctive boomerang marking on its Sopwith Camels. Illustrated is B7406 of "C" Flight as it appeared in March 1918. This aircraft was delivered to the Squadron on 25th February 1918 by Et G. Jones (now Air Marshal Sir George Jones, KBE, CB, DFC), and was still in service three months later.

The Australian Prime Minister, the Hon William Morris Hughes, tries out the cockpit of a No 3 Squadron RE8 in July 1918.

RE8, Herbert could do nothing but concentrate completely on keeping it in the air. Meanwhile, his observer, Lt F. A. Sewell, kept his mind on other things by blazing away with his Lewis gun at anything that moved in the mist, secure in the knowledge that it had to be enemy! Finally, the inevitable happened and Herbert was forced to fly right through the uppermost branches of a clump of trees, the result of which was a stick jammed in the starboard aileron, making the aircraft difficult to control and further flight near the ground suicidal. Herbert then forced the machine up through the fog toward the clear air above, while Sewell climbed on to the top wing and grimly edged his way to a position where his weight was sufficient to overcome the effect of the immovable aileron. Free at last from the intense concentration of the previous half-hour, Herbert was able to work the branch loose by continuous movements of the control stick, enabling Sewell to regain his cockpit for the long flight home. Possibly the observer spent the time ruminating on the fact that he had perched himself on a

The sparse comforts of a pilot's quarters in France.

Lts R. C. Armstrong (pilot) and F. J. Mart, responsible for the capture of a German Halberstadt, smile happily from their RE8. Right. Another picture of Armstrong and Mart ready for a bombing mission. The nearest bomb bears the inscription "From the Aussies."

wing which projected more than ten feet beyond its supporting interplane struts, an overhang that sometimes after a heavy landing, drooped down like the ears of a dejected spaniel.

As the German offensive ground to a halt, the opposing infantry forces began to consolidate their trench systems on the Somme, with a resultant increase in the work demanded of No 3. There was little change from the seemingly endless routine until 9th June. Late in the morning the normal work of the Australian base, now at Flesselles, was upset by the unexpected and hurried landing of a German Halberstadt two-seater, the crew of which were promptly taken prisoner at revolver point. The reason for this gift was made apparent immediately with the almost simultaneous arrival of a squadron RE8 crewed by Lts R. C. Armstrong and F. J. Mart. The excited pair had spotted the German machine heading for home, and Armstrong headed it off, spoiling for a fight. The German crew were lost and inexperienced, and wisely showed no inclination to incur the wrath of the Australians by returning their opening shots. Instead they allowed themselves to be turned back in a westerly direction towards captivity, nervously aware that the Vickers gun of the aircraft 100 yards behind was constantly trained on them, as were the goggled eyes of the two men. In this fashion they arrived at the Australian aerodrome. The Halberstadt, completely undamaged, was handed over later to the Australian Government as yet another war trophy.

Throughout the rest of June the work of No 3 Squadron steadily increased in preparation for the planned assault on the German strongpoint at Hamel, which culminated early in July in some of the most intensive and successful work that the squadron had ever achieved. At this time Lt Mart again figured in adding a further victory to the squadron record when he and his pilot, Lt D. F. Dimsey, went to the aid of another RE8 and shot down a Pfalz from a formation of enemy fighters. Air combats were frequent and on occasions were actually sought by the Australians when freed from their main duties, despite the limitations of their machines for this type of offensive warfare. During this period the squadron was also instrumental in providing a new form of assistance to the ground troops, when it was given the job of adapting a German idea to supply ammunition to forward troops in isolated positions. A practical system involving the use of modified bomb racks and parachutes was evolved, assembled and tested satisfactorily by the efforts of Capt Wackett (who brought his inventive genius with him when transferred from No 1 Squadron in Egypt), together with Sgt Nicholson and the squadron mechanics. The RE8's of No 9 Squadron RAF, operating from the same aerodrome, used this system to supply ground forces during the Hamel attack, providing a further example of the ever-increasing value of the air force units in the developing field of close support.

Officers and men of No 3 Squadron inspect the undamaged Halberstadt after its capture on their own aerodrome. Right. The receipt issued to No 3 Squadron for the maps and personal effects of the Halberstadt crew.

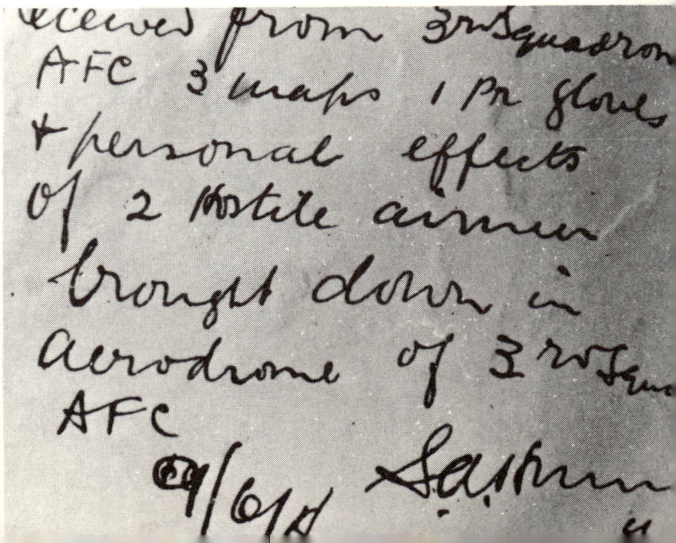
eceived from 3rd Squadron
AFC 3 maps 1 pr gloves
& personal effects
of 2 hostile airmen
brought down in
aerodrome of 3rd Squ
AFC
9/6/18

Another view of the captured German machine showing its fine lines to advantage.

With the success of the Hamel battle, the squadron maintained its routine work for the rest of July, making the enemy ground forces as uncomfortable as possible and collecting the information required for the forthcoming August offensive. It was on the first day of that month when the "balloonatic" pilot of a German Albatros fighter destroyed four British observation balloons and then somewhat unwisely chose his homeward path through an area of sky containing three Australian RE8's. His apparent supposition that the observation machines would not impede his passage proved incorrect, but he managed to avoid the fire from the first two, only to become the victim of Lt H. B. James, the observer of the third machine flown by Lt A. E. Grigson. (If he was fortunate enough to survive his resultant crash, the German pilot's mortification may have been softened had he known that he had been bested by a team who already had notched up several victories.)

Bad weather during the opening of the Allied offensive on 8th August restricted aerial activity, but as soon as conditions moderated sufficiently the airmen of both sides threw themselves into action. Naturally, No 3 was chiefly concerned with the ever-changing events on the ground, with fighter formations providing the necessary protection to enable them to concentrate fully on their job. Enemy machines which slipped through the cordon had the choice of taking out their spite on either the advancing infantry or the observation aircraft. In view of the increased opposition the squadron losses were surprisingly light and, from a purely impersonal viewpoint, almost negligible when set against the great amount of damage they inflicted on the enemy. By now the infantry commanders were fully aware of the value of co-operation with the corps squadrons, and their numerous requests for assistance kept the machines constantly in the air. That this service was never lacking is a great tribute not only to the pilots and observers, but also to the mechanics, riggers and armourers whose efforts made it possible.

During the last week of August the British forces began to push their way southward out of the Somme area, and by the end of the month there were positive indications that the Germans were in retreat. Bad weather again hampered the observation of enemy movements and the position of the friendly troops, and accurate reports required daring and

Personnel of "B" Flight of No 3 Squadron pose in front of an RE8 in Belgium in 1918.

Lt J. J. Pengilley in the cockpit of a well-worn RE8. To lessen the risk of fire in a crash, some pilots filled the reserve fuel tank behind the engine with water or fire extinguisher fluid.

skill such as that displayed by Lts J. J. Pengilly and O. G. Witcomb when they flew their RE8 low along the Somme to mark any remaining crossings. To get double value for their money they also noted down for future attention the various machine-gun nests that disclosed their positions by opening fire, with Pengilly weaving his machine through the bursts like a quail darting along a line of shooters. On the following day Lts G. E. Kilburn and W. P. Moore were carrying out the same type of mission when they were jumped by a squadron of enemy fighters which suddenly materialised from the cover of low cloud. It says much for the ability of the crew and the RE8 that they fought long enough to return the enemy fire with 200 rounds of their own before being overwhelmed. Fortunately, although the engine was disabled, the uninjured Kilburn made a successful emergency landing in safe territory, where he and his slightly wounded companion were rescued by Australian infantry.

The opening weeks of September 1918 saw the German army in full retreat toward the Hindenburg Line, harried by the airmen in the same way that the Middle East squadrons were to attack the fleeing Turks later in the month. However, the German retreat was still relatively orderly with a view to setting up resistance along that last line of defence and there were still large numbers of enemy aircraft in the sky. These were unable to put No 3 out of business, but the weather almost succeeded in doing so about the middle of the month when an early morning gale wrecked the hangars and badly damaged most of the aircraft. With an attack scheduled for the next day the situation was desperate, but despite the appalling number of problems the squadron was operational again by the same evening due to the mighty efforts of the aerodrome staff under Capt Ross, the equipment officer.

With the enemy entrenching himself along the line from which he had begun his March offensive, the Australians settled down to the familiar routine of gathering evidence for use against him during the

Mechanics of No 3 Squadron overhauling an aircraft of "B" Flight at Premont in France in October 1918.

Men of No 3 Squadron with a veteran RE8 in Belgium. The serial number H7042 indicates that this aircraft was an Aircraft Repair Depot rebuild.

final assault at the end of September. During the initial stages of this attack, aircraft of the squadron were called upon to drop supplies to isolated American units surrounded by the enemy in a miniscule rehearsal for events of a quarter-century later when the Americans would return the favour.

Once again the weather was so foul that the observation machines were forced to carry out their work at dangerously low altitudes, well within the range of the numerous machine-gun nests, but despite this casualties were surprisingly light. By the end of the first week in October the Germans were definitely on the run and the work of the squadron began to lessen until eventually patrols were being flown only on request from the ground commanders. Several Bristol Fighters had been allotted to the squadron about this time for use on long-distance reconnaissance, and these were formed into a special flight under Capt E. J. Jones to keep an eye on the general situation until there was nothing further to report.

With the coming of the Armistice, No 3 was retained in Belgium to operate an aerial postal service for the remaining forces; a rather prosaic finish to the operational story of a unit which had been described as the best corps squadron in France. With final disbandment in February 1919, the surviving members returned to Australia in May, leaving behind them a proud record of achievement which ranks with the best in Australian aviation history.

The wireless section of No 3 Squadron at Tarcienne in Belgium shortly before disbandment in February 1919.

Early training in England before the establishment of the AFC Training Wing was carried out on a variety of types such as this RE7, serial 2357, at Harlaxton. Right. An Avro 504J, serial A5927 and a Nieuport 12 dispute right of way in one the less serious accidents common to training airfields.

The Bristol Fighter also had its role to play in the advanced training field, as seen here at Minchinhampton in 1918. Right. The Armstrong-Whitworth FK3 was a less well-known type which served usefully as a trainer for AFC pilots in 1917.

To help overcome the potentially dangerous instability of the RE8, some training aircraft were fitted with the enlarged fin seen on this machine at the AFC base at Leighterton. Right. Also at Leighterton was this Bristol Fighter, with Capt Wrigley of No 3 Squadron AFC standing in front.

Some of the DH5 aircraft with which No 2 Squadron AFC began operations were donated by public subscription. A9242, seen at Harlaxton during final training, was presented by the women of New South Wales.

CHAPTER 4 THE FIRST FIGHTER UNIT

Within a fortnight of No 3 Squadron arriving in France, the second Australian squadron crossed the English Channel to land at St Omer on 21st September. The safe completion of the flight from Harlaxton by all its machines in one day created a record and, as a result, the complete squadron arrived at Warloy near Baizieux on the following day, ready to carry out their duties as the first fighter unit of the Australian Flying Corps.

Commanded by Maj Watt, who had been placed in charge when the squadron was first formed in Egypt, its equipment consisted of the DH5, a pugnacious-looking scout whose appearance belied its actual quality for air-fighting at altitude. One of its less endearing features was the fact that if its rotary engine shed a valve tappet rod (a quite common occurrence, which was eventually overcome by the fitting of a simple clamp), this rod would cut through the thin engine cowling like a demented tin-opener before coming up against the relatively immovable mass of the Vickers gun, which projected forward above the engine. This solid resistance was sometimes sufficient to wrench the motor right out of its mountings, with disastrous results for the poor pilot. Also, the DH5 had earned a reputation, not completely justified, for being difficult to handle at low speeds. When considered with the fact that its performance fell off rapidly above 10,000 feet, it can be appreciated why the first operations of the squadron brought meagre results.

There was no lack of aggression on the part of the Australian pilots, but their job was certainly not easy when enemy two-seaters could outdistance them and the German fighters could dictate the terms of any combat. As a result, the squadron suffered its first casualties during October without being able to even the score. Lt I. C. F. Agnew was the first to be lost when he was forced down and captured, thereby achieving the dubious honour of becoming the first casualty of the AFC as a result of combat in France. Shortly afterwards Lt D. G. Morrison died from wounds received when a pack of Albatros scouts jumped him while he was separated from his formation.

This galling state of affairs continued into November when the role of No 2 was altered in preparation for the forthcoming Battle of Cambrai. Among the advantages of the DH5 were its strength and ideal forward view for the pilot and, despite the fact that it was armed with only one gun, its low-level performance made it a logical choice by the authorities for close support ground-attack work. Thus the second week of November found the squadron busily engaged in practice for this hazardous work.

The 20th November 1917 saw the Australians applying their new skills in deadly earnest as they maintained continuous supporting patrols for the infantry and tanks which opened the offensive. Thick fog added to the dangers associated with the low level bombing and machine-gunning of the enemy, making it impossible to use large formations. Instead, the pilots split up into pairs after reaching the designated area, then hunted their targets virtually at ground level. Under these unfavourable conditions a price had to be paid!

Capt J. Bell, who led the first flight over the lines just after dawn, was mortally hit by rifle fire. Lt L. N. Ward was shot down and captured. Another pilot was wounded. More than one third of the squadron aircraft were shot down, fortunately without further loss of personnel, and the battered remainder bore mute testimony to the accuracy and intensity of the

Left. The training period for No 2 Squadron personnel had its share of mishaps. A serious crash involving RE7, serial 2402. Above. The spectacular result of an argument between Armstrong-Whitworth FK3, serial A1479, and a tree.

ground fire. It may have been this day that earned Lt L. H. Holden his nicknames of "Lucky Len" and "Homing Pigeon", for it was inconceivable that his DH5—literally a flying wreck—could absorb so many German bullets and still remain airborne. Even more miraculous, he himself was not hit! (During the German offensive of the following March, Holden is said to have used up an average of one aircraft per day in similar action, which can be taken as a tribute to his determination.)

This was the pattern of work for the squadron during the next fortnight, with several more inevitable losses and much damage to the aircraft. However, the appearance of increasing numbers of German aircraft gave the Australian pilots the opportunity to prove themselves in the role for which they had originally been trained. In order to come to grips, the German pilots now had to sacrifice their advantage of height and meet the DH5's on more even terms. As a result of this, the first victory was recorded on 22nd November, when Lt F. G. Huxley surprised an Albatros as he climbed steeply away from the ground after bombing and strafing an infantry column. Huxley's instant reaction and marksmanship combined to send the enemy scout crashing to the ground before the German pilot knew what had hit him. On the same morning Lt R. W. Howard, in company with some RFC scouts, was instrumental in forcing a DFW two-seater to land in the British lines. Further successes were achieved during the final days of the month, but their main duty was still the harassment of the enemy ground forces.

On 30th November the Germans launched a surprise counter-attack assisted by large numbers of aircraft, and for a time the Australian airmen had to restrict their own ground attacks and concentrate instead on the enemy machines. These were mainly two-seaters which were dangerous enough at any time, but particularly so when near the ground and

An exceedingly rare bird was this Rolls-Royce Hawk engined BE2e, seen in flight over Cranwell in November 1917.

Another presentation DH5, "The Upper Hunter", serial A9197, seen at Harlaxton with Capt Stan Muir at the controls.

thus difficult to attack from the blind spot underneath. Despite this, several fell victim to the squadron pilots during the next week, with Huxley downing a DFW in flames on 6th December as the final tribute to the gods of war by the squadron DH5's. This type was then replaced by the SE5a—undoubtedly one of the best scouts produced by either side during the Great War.

Re-equipment and training kept the squadron out of operations until the latter half of January 1918 when they took up position at Savy, the original home of No 3 Squadron. From here they at last began to fly as a true fighter unit, searching the skies at the high altitudes where their new aircraft operated most efficiently. This meant patrols in excess of 15,000 feet, far above the smoke and fumes of the battlefield through which they had previously flown, and concerned only with enemy aircraft and the relatively predictable anti-aircraft fire rather than the invisible curtain of steel encountered near the ground. To offset these advantages, they had to fly in open cockpits with no form of heating and no oxygen supply, luxuries which were later deemed essential for sustained high altitude work.

The opportunity for the No 2 Squadron pilots to test their new equipment in actual combat was delayed until mid-February by the weather and the reluctance of the enemy to come to grips, but 18th February gave them their first chance. Huxley again played a leading part when he and Lt A. L. Paxton encountered a flight of Albatros scouts, and Huxley sent their leader spinning earthward, upon which the others judiciously made for home. The two Australians followed up this action by forcing down a two-seater. A little later in the same day another pair, Lts Howard and Holden, achieved similar results. On succeeding days when weather permitted, the Australians went out in pairs and small groups, but despite their numerical superiority the German formations usually sheered off after the opening skirmish, preferring to hold themselves in readiness for their forthcoming offensive in March. However, on 12th March the squadron went out in force and turned the tables. Ten SE's surprised half a dozen Albatros scouts which were keeping a protective eye on a reconnaissance two-seater, and from the resultant dogfight Holden sent down one of the enemy fighters in flames for a very decisive victory. Leaving the alley-fighting to his companions, Lt R. W. McKenzie carried straight on to the most important objective and very efficiently disposed of the reconnaissance machine.

Skirmishes of this nature took place until the German attack was launched on 21st March when the enemy air forces threw away their caution and came out in a determined effort to rule the sky. This was certainly no time to go roaming singly or in small numbers, and the Australians maintained large patrols which achieved a number of victories during the opening days. Unfortunately, these successes were offset by the loss of Capt Howard, by then a flight commander, who was fatally wounded on the 22nd, but he was avenged several times over on the following day. Throughout the remainder of the month the mounting toll of aircraft destroyed by the Australian and Allied squadrons played an important part in finally halting the German advance by denying the enemy the observation and close support facilities so essential for success.

However, the initial advances had necessitated several moves for No 2, and early April found them operating from Bellevue aerodrome in conjunction with the Sopwith Camels of two RAF squadrons (the Royal Flying Corps and the Royal Naval Air Service had been amalgamated on the first day of that month to form the Royal Air Force). From here, on 4th April, with the squadron still settling into its new quarters, Lt L. J. Primrose celebrated the house-warming by shooting down an Albatros which he and Lt G. R. Brettingham-Moore encountered over Villers-Bretonneux. This victory was followed by the destruction of an observation machine two days later, but then the weather became a greater menace than the enemy aircraft. Regular patrols were impossible, and on many occasions when aircraft were able to take off the weather closed in again before they returned, resulting in forced landings all over the countryside.

As soon as the bad weather eased, sweep formations were introduced. This idea was derived from the German practice of combining four of their

Pilots and aircraft of No 2 Squadron AFC at Harlaxton photographed immediately before leaving for France on 21st September 1917.

As though pensively considering their forthcoming journey to France, a fitter and his charge stand quietly together at Harlaxton, with a DH4 day-bomber in the background.

most efficient squadrons into a mobile fighting group, to be used at any point on the front which required extra attention. The British system as it effected No 2 Squadron was to have the SE5a unit operating at 16,000 feet as top cover for the RAF Camels. These latter were at 12,000 feet and 14,000 feet where their outstanding manoeuvrability could be exploited to the full. This worked quite well, perhaps too well for the pilots since the enemy aircraft made it their business to shy clear of the large formations, leaving the Australians with the feeling that they were little more than aerial policemen patrolling a beat while the criminal element hid in the alleys and dark doorways of the clouds waiting for them to pass. Had they realised that the Germans were once again reserving their strength for a renewed offensive in early June their impatience might have been easier to bear, although several engagements during May helped to relieve the monotony.

These combats resulted in several more successes for the squadron, including a triplane sent down out of control by the CO, Maj Murray Jones, who still displayed the same fighting spirit he showed with No 1 in Egypt. Although an out-of-control report was not necessarily a victory because pilots sometimes used this method to escape from a nasty situation, there was no ruse about the way Lt E. D. Cummings left one of these encounters early in May. Having just put paid to a lone triplane, he was jumped by four more of the same type, who angrily proceeded to riddle his SE with bullets. With petrol and oil spraying from smashed pipes, the instrument panel in front of him wrecked, and no elevator control, his aircraft began to spin uncontrollably toward the ground a mile below. By some miracle unwounded, Cummings wrestled furiously with his machine until just before impact, when he succeeded in getting the nose up into an attitude loosely resembling level flight. He still crashed with sufficient force to break his safety belt and be thrown out of the wreck, but was relatively unhurt when rescued from no-man's-land by Australian infantry.

In the latter half of May the enemy renewed his activities, but the weather again took a hand, and the squadron's work was interrupted for the rest of the month. Fine weather on the first two days of June saw formations from both sides out in force and, in retaliation for the inactivity of the past fortnight, No 2 accounted for almost the numerical equivalent of an enemy squadron in just two days. On the debit side they lost Lt A. R. Rackett who was shot down and taken prisoner.

Following the German pattern, the group was then ordered to Foquorolles in anticipation of an attack in that area and, when it came on 6th June, their immediate task was the hateful ground attack work. The scale of operations was nowhere near

SE5a aircraft of No 2 Squadron AFC wait at an advanced landing ground near Bapaume early in 1918. "C" is probably A8934, one of the first produced, and by this time a real veteran.

This scene of an SE5a warming up at the AFC training base Minchinhampton late in 1918 could be a flashback of twelve months to the time when No 2 Squadron first received the type.

as large as that encountered on their previous missions of this type and there were no losses during the four or five days they were thus occupied, but it was still a relief when they resumed their normal offensive patrols. On the 11th the first victory in their new sector was chalked up by Lt T. J. Hammond, who picked out and destroyed a two-seater from a formation of six. On the following morning the same pilot himself became the victim during an engagement with a pair of Pfalz scouts which attempted to dive on the middle Camel formation. Capt R. L. Manuel of No 2 added one of them to his mounting score, but Hammond was lost when the remainder of the German formation joined in. Unfortunately for the enemy, they were now down to the ideal level for the Camels to show their merit, which they promptly did by shooting down three more. As further recompense for the loss of the Australian, Capt R. C. Phillips later in the day added four more victories on his own account; all achieved in the course of one patrol and no mean feat by any standards. His first victims were two Fokker triplanes from a formation of six, immediately followed by an LVG two-seater, one of a pair trying to slip quietly past the commotion above them. Later during the sweep Phillips picked off a Fokker DVII from a formation of 10 which tried to get past the Australians to attack the Camels below. (The Fokker DVII was now being encountered in ever-increasing numbers, and the performance of this new type lessened the superiority previously held by the British fighters. The SE5a was one of the few aircraft which stood an even chance against it at high altitudes.)

Bad weather again curtailed flying and there was no further significant action in this sector before

SE5a, serial C9539, of "C" Flight leader Capt H. G. Forrest comes to grief on Savy aerodrome in France in March 1918. Forrest was very successful against the enemy in this machine during the opening days of the German offensive at this time.

SE5a machines of "B" Flight lined up on the No 2 Squadron aerodrome at Savy on 25th March 1918.

the group moved north on 20th June. This move finally resulted in the transfer of the Australian squadron to the 80th Wing, where they joined their compatriots from No 4 squadron at Reclinghem. The association was marked by friendly rivalry between the two units to see which squadron could inflict the most damage on the enemy, as a result of which, fortunately, the only ones to suffer were the Germans. Freed from the restrictions of the large sweep patrols, the SE's combined with the aircraft of No 4 in a more flexible routine, while still maintaining the logical pattern of top cover for the slower Camels. In addition they tried to out-do each other in the amount of damage inflicted on enemy ground installations, with dangerous attacks on observation balloons thrown in for good measure.

No 2's first victory in this new locality came on 4th July when Lt F. R. Smith destroyed a Fokker DVII. Several days later the squadron made an attack on the railway yards at Lille, with Murray Jones leading the formation of 10 off the ground at first light. Finding their way to the objective by compass course through bad visibility, they launched a damaging low-level attack on trains and railway installations, leaving a few of their number "upstairs" to guard against interference from enemy fighters. Trouble was not forthcoming, but the success of the raid was marred by a thickening fog, which caused half of the pilots to lose their way and suffer forced landings, all fortunately in home territory. Unlike their previous ground attack work, carried out during the frenzy of an offensive against bodies of troops and gun emplacements, these raids against major installations were the result of careful pre-observation and planning by the Australian squadrons and became a feature of the work of the following months.

This did not alter the fact that they were primarily airfighters, and there were plenty of targets for the high-flying SE's due to the large number of aircraft maintained in this northern sector by the Germans as insurance against a penetration of their right

No 2 Squadron officers photographed at Baizeux in France. Right. Officers of No 2 Squadron at Minchinhampton after service overseas. From left (standing) are Capt L. H. Holden, Maj R. C. Phillips, Capt W. A. Robertson; (kneeling) Lt H. Clark, Col Oswald Watt, and Lt R. L. Clark.

Capt L. H. Holden and friends pose before his SE5a at Minchinhampton late in 1918. Holden was one of the instructors in the AFC Training Wing who gave their fellow-Australians the benefit of their combat experience.

flank. Despite bad weather during the remainder of July, No 2 scored several victories without loss, although Cummings again had a narrow escape on the last day of the month. After destroying an LVG two-seater, he was heading for home when he encountered four more of the same type. Since his previous victim had been one of a formation of eight, this group must have seemed relatively easy pickings and he wasted no time in diving on his selected target. This dive was almost his last, because he practically collided with the German, and both aircraft dropped out of control. Cummings regained control but his badly hit opponent continued to spin earthwards and disintegrated.

The month of August and the number of enemy aircraft brought a reversion to the sweep system of operations, with No 2 alternating with the Bristol Fighters of 88 Squadron RAF in providing top and middle cover for the low-flying Camels of No 4. As before, the size of these formations acted as a very efficient deterrent to the enemy fighters, but there was still the occasional high-flying German two-seater trying to slip past the upper formation at maximum altitude. This was not such a suicidal action as it may seem. The high-altitude performance of a two-seater such as the German Rumpler enabled it to maintain a relatively straight course for home while its attacker, tenuously balanced like a tight-rope walker in the thin atmosphere at 20,000 feet, could easily lose several hundred feet in any sudden manoeuvre. Once lost, this altitude had to be regained slowly and carefully and this type of stalking required steady control and plenty of patience. Lts J. J. Wellwood, C. R. Ebeling and V. E. Knight met one of these aircraft at 19,000 feet on 1st August, but the German made their job slightly

No 2 Squadron SE5a aircraft on their aerodrome near Lille in November 1918.

Mission completed! An SE5a taxis in after landing near Lille in November 1918.

easier by diving for home, a dive which continued right into the ground after a long burst from Wellwood at close range. Other victories were recorded against enemy fighters attempting to disrupt the ground attack work of the Camels, over whom the SE pilots still maintained a brotherly watch.

In mid-August No 2 took part in devastating raids on enemy aerodromes on the outskirts of Lille. These attacks on Habourdin and Lomme were made on the 16th and 17th respectively, with the two Australian squadrons bombing and strafing under the protective cover of RAF Bristol Fighter and SE5a units. Considerable damage was caused and the success made even greater by the fact that No 2 suffered no losses, the only casualty being a pilot from No 4. The resultant effect on the air opposition in this area was very noticeable, and aerial combats throughout the rest of the month were spasmodic, though no less fierce. With the Allies beginning to throw their weight against the German lines in the south, the enemy was now forced to concentrate as much of his air power as possible in this area, and by ranging in this direction the Australians became involved in some heavy action.

The general retreat of the enemy forces became apparent during September, and his air units found it impossible to provide overall protection. Even in the areas where maximum concentration was deemed necessary, replacement and supply shortages made his task even more difficult. However, there was no lack of determination on the part of the German pilots and, as the front narrowed and the enemy squadrons became more concentrated, they fought hard as if denying the impending defeat that was now becoming more and more apparent. No 2 engaged in some very torrid actions during this period with excellent results, their own losses being very small. To further bedevil the enemy, they abandoned their high-flying duties on occasions during October to take part in ground attack missions similar to those carried out against Habourdin and Lomme, this time concentrating on Lille and Tournai. Once again aerodrome and railway installations were heavily hit, together with any targets of opportunity worthy of bombs or bullets. These raids were carried out without loss, but Lt A. L. Long only just managed to regain his own lines after his petrol tank was holed by a fragment from one of his own bombs. Capt E. E. Davies was similarly lucky after his aircraft was damaged by ground fire.

The last few weeks of the war were notable for the large numbers of enemy aircraft encountered, and fighting was frequent and heavy. Often outnumbered, the Australians were always ready to tackle the enemy formations, at times under highly unfavourable conditions, as illustrated by an action involving Capts E. R. Dibbs and G. H. Blaxland. After a long battle against a pack of Fokkers, one of which had fallen in flames to Blaxland, the pair were returning home together when seven more

Replacement pilots for the SE5a squadron trained on their operational type at Minchinhampton, where these aircraft were being serviced.

SE5a machines D/379 and E5765 of "A" Flight being serviced by mechanics of No 2 Squadron near Lille in November 1918.

Fokkers hove into view above them, obviously unaware of the two Australian machines. Blaxland, with both guns jammed, was undoubtedly keeping his fingers crossed that none of the German pilots would glance in their direction, when Dibbs, with only one gun in action, suddenly pulled up the nose of his aircraft and let fly at the formation. Fortunately, the ability of the SE to withstand a full-throttle vertical dive saved the escaping pair, but whether Dibbs withstood his companion's blistering remarks with the same success after their return has not been recorded!

The successes achieved by the squadron in these final battles again far outweighed the few losses suffered, but the week leading to the Armistice found them with a lack of targets, many of the German squadrons having insufficient fuel to get their aircraft into action. The result was that No 2 finished the war almost as it began, withdrawn from its true element, but now harassing the thoroughly disorganised enemy troops fleeing eastwards toward Germany. The squadron's last casualty occurred during these operations, when Capt F. R. Smith was brought down on 9th November by ground fire. Successfully evading the German troops in the immediate area, he eventually obtained some civilian attire, and after several days of observing at first hand the terrible sights of a beaten army (which included seeing a squad of soldiers kick their NCO to death), he set out, still undetected, to walk the 40 miles back to his unit.

The Armistice celebrations were still in full swing when he walked into the squadron Mess, and after a rousing welcome, he too wasted no time in joining in the festivities and toasting the outstanding success of Australia's first fighter unit.

During a visit to the AFC Training Wing, Gen Birdwood chats to Maj R. C. Phillips. Col Watt stands behind the general with Maj R. S. Brown at right and Capt Tooth obscured by Maj Phillips. Right. Officers of Group HQ 1st Wing at Tetbury. Back Row:— Lt Chambers, Lt Cooke, Capt Morgan, Lt R. O. C. Matthews, Lt J. A. Robertson, Lt E. L. Spencer, Lt Marsden. Front row:— Capt Brake, Capt E. N. Tooth, Col Watt, Capt G. C. Matthews.

Fitters tinker with a cold engine to enable J. Sands to take off for gunnery practice at Minchinhampton. Right. Even at rest on the ground the aggressive-looking SE5a seemed to sniff the air for a scent of the enemy.

Preparing an SE5a for a training flight. The pilot, J. R. Gordon, destroyed at least five enemy aircraft as an observer with an RFC Bristol Fighter squadron before transferring to the AFC for pilot training. Right. Gen Birdwood (rear cockpit) about to be given a demonstration flight at Leighterton.

Camel C127 displays an Australian emu emblem on fuselage which, together with a similar kangaroo emblem, distinguished aircraft of the AFC Training Wing. Right. Also at Minchinhampton was this Bristol Fighter fitted with a Sunbeam Arab engine.

SOPWITH SNIPE

In October 1918, No 4 Squadron was re-equipped with the Sopwith Snipe, and used this type to great advantage during the last weeks of the war. E8198 displays the later-type white vertical bar adopted as the unit marking.

SOPWITH PUP

© CHARLES SCHAEDEL

Apart from the four operational units, Australia maintained four training squadrons in England equipped with a variety of aircraft. The Sopwith Pup was a particular favourite, and B1764 bears the kangaroo marking, showing it to be a machine of the Australian Training Wing.

Sopwith Camels of No 4 Squadron AFC lined up on Bruay aerodrome on 26th March 1918. Shortly after this the boomerang squadron marking was replaced by a vertical white bar behind the cockpit.

CHAPTER 5 LAST BUT NOT LEAST

Last but not least of the Australian Flying Corps squadrons, No 4 was formed in Australia and arrived in England in March 1917, under Capt A. Lang. During its training period in England the command of the squadron was handed to Maj Sheldon, one of the officers who had gained experience with No 1 in Egypt. On departure for France on 18th December 1917, it was led by Maj W. A. McCloughry, who had seen action in France with No 2 Squadron AFC, as had Capt G. C. Matthews, who was posted as a flight-commander in the new squadron. His fellow flight-commanders were Capts O'Hara Wood and Hughes-Chamberlain, the former an Australian. Both these men were RFC officers with previous combat experience.

The squadron was equipped with the Sopwith Camel, a type which possessed outstanding aerobatic qualities, although it was not exceptionally fast. Despite some dangerous tendencies which an experienced pilot could use to advantage in combat, the value of the Camel may be judged by the fact that the type recorded more victories than any other Allied aircraft used in World War I.

Like No 2, the squadron crossed the Channel to St Omer without mishap. From here they were sent to Bruay, where they spent several weeks learning their assigned sector. During this training the squadron suffered its first casualties, when Lts A. M. Anderson, R. H. Curtis and J. N. Cash were killed as the result of a mid-air collision during practice formation flying on 6th January. A week later, shortly after operations were commenced, Lt F. B. Willmott became the first casualty over the lines when engine failure forced him into captivity. Bad weather during this period restricted activity, but the 24th January saw the squadron open its account on the credit side when a DFW two-seater fell to the guns of Capt O'Hara Wood.

However, the enemy squadrons were husbanding their resources in preparation for the March offensive, and combats were few and largely indecisive. It was early February before further victories were added, when O'Hara Wood took Lts E. F. Pflaum and A. H. Cobby out on the 3rd in a hunt for trouble. They found it in the form of an equal number of DFW's, and each of the Camel pilots notched up a score. Cobby's victim was the first of many which gave him the honour of being the top-scoring pilot of the Australian Flying Corps.

Activity increased in the latter half of the month, but the squadron suffered its first fatality in action on the 17th, when Lt C. H. Martin's Camel inexplicably broke up while on patrol. The score was evened with the destruction of a two-seater by Lt F. S. Woolhouse, one of the unlucky pilot's companions. Spasmodic aerial encounters were interspersed with ground attack work, and during this period two more pilots were taken prisoner, Lt A. Couston being forced down in an air fight and Lt W. B. Randell as the result of ground fire. The latter was particularly lucky to survive. After a fruitless offensive patrol Randell followed instructions to use up his ammunition on suitable ground targets on the way home, but after shooting up several machine gun emplacements he found himself flying directly towards an enemy observation balloon. He immediately attacked it, but heavy defensive fire put his engine out of action and he was forced to land in a field close to the scene of his previous attack. Naturally, the German survivors were not especially well disposed towards him, and bayonets were being flashed in close proximity to his throat when an officer rescued him from further beating and took him into captivity. (Perhaps at this stage Randell regretted the fact that he had escaped a court-martial in England during his training period. He had been told by a senior officer to try his hand at an RE8, but being short in height found that he could only just reach the rudder bar by stretching his legs at full length. Having been informed of the necessity to keep a tight rein on this type at all times, and now aware that he would be unable to do so, Randell immediately climbed out of the cockpit and made some very Australian suggestions as to what

View from the cockpit of a Camel. The tube between the guns is the Aldis telescopic sight, with ring and bead sights in addition. The windscreen side shields were often removed to increase forward visibility.

the officer could do with his so-and-so RE8. The threatened courtmartial was waived in favour of sending him to France on the smaller Camel.)

These air and ground operations were but a rehearsal for the heavy work to come for No 4 during the German push of 21st March. This offensive was preceded by increased enemy reconnaissance activity, with large numbers of German fighters providing protection for the important observation machines. At this time the fighter squadrons opposite the Australians included the roving group commanded by Baron Manfred von Richthofen, and the first encounter came on 16th March. Seven Camels had just completed a bombing raid on the Douai railway installations when 16 Albatros scouts of the "Red Baron's" group arrived on the scene. In the ensuing free-for-all the Australians acquitted themselves well, and there were two, and probably three, empty chairs in the German mess that night. The pilots responsible were Lts G. F. Malley, C. M. Feez and A. W. Adams. Missing from the Australian formation after the fight was Lt W. H. Nicholls who had only been with them a little over three weeks. He was forced down to ground level, just failed to make it safely into home territory, and was captured by German front-line troops.

After another slack period due to the absence of enemy aircraft and the foggy weather, the squadron again had a brush with a large formation thought to be composed of the Baron's aerial henchmen. Having the advantage of height the 10 Camels immediately attacked and, before the dogfight broke up in the ground mist, four enemy machines had been accounted for. Two went down to Cobby, and Lts Pflaum and A. E. Robertson each out-fought another. The only Australian casualty was Lt G. M. Elwyn who was forced down unhurt behind the British lines.

Then the storm broke! Out of the fog which obscured the sunrise on 21st March, hordes of grey-clad German infantry men advanced on the British positions, and No 4 found itself part of the frenzied attempt to stem the tide. The following days became a succession of ground level attacks through mist, dirt and smoke-filled air, with the Australians bombing and machine-gunning anything in sight. In return they were fired on from the ground and occasionally from the same level when German aircraft managed to evade the protecting screen of scouts. These intruders provided several more additions to the squadron scoreboard, but their own losses during this hectic period were light, amounting to one killed and three wounded.

Not an uncommon sight on training airfields from which Sopwith Camels operated, as pupils learnt this type's tricky habits the hard way.

Pilots with operational experience were posted to the AFC Training Wing to train their fellow-Australians. This all-white Camel was the personal machine of Capt G. F. Malley who had earned distinction with No 4 Squadron.

Within a week the Germans had been halted in the area covered by No 4, and the squadron returned to its normal offensive patrols. Several more victories were recorded, including two triplanes brought down during one sortie by Lt E. R. Jeffree, who had been with the squadron only three weeks. But the following day saw the capture of Lt Feez resulting from the failure of his engine.

Regular bombing and strafing patrols were still carried out in conjunction with the search for enemy aircraft, but a sudden break-through by the Germans early in April again saw the squadron throwing its full weight against the attacking ground troops. As before, there was no rest for men or machines during a week of maximum effort, but apart from the large amount of damage to the aircraft which had to be accepted as a normal result of this work, casualties were again light. Lt F. S. Woolhouse, one of the squadron's old-stagers, lost his life, and Lt H. K. Love, who had only just been posted to the unit, was brought down and captured.

With the enemy advance finally halted, No 4 returned to its interrupted task of planned bombing raids and offensive patrols until the end of April, when the attentions of the German bombers and long-range guns finally rendered their aerodrome at Bruay untenable and they were forced to move to Clairmarais. Their successes in the air during May became more frequent as the German airmen strove to relieve their ground troops of the constant pressure being applied by the Allied squadrons. No 4's own losses from combat were small, but a large amount of material damage was incurred about the middle of the month when thick fog closed in while two flights were taking part in a raid on Armentieres. After fighting their way home through a large pack of defending fighters, the British and Australian airmen found it impossible to locate their bases and soon crash reports were coming in from all over the country. Miraculously, the only casualty suffered by the Australian squadron was Lt F. W. Webster of the RAF, who received a broken foot to

Pilots of No 4 Squadron at Clairmarais in May 1918. From left:—Lts J. H. Weingarth, R. G. Smallwood, R. C. Nelson, W. S. Martin and A. T. Heller. Heller, who had only joined the squadron on the previous day, eventually became a flight commander in the unit.

Pilots of No 4 Squadron's "C" Flight at Clairmarais in France, June 1918. The later vertical white bar squadron marking is visible on the aircraft in the background.

terminate his short period of attachment to No 4. However, six Camels were written off and, together with the loss of Lt O. C. Barry to the German fighters, this marred what was otherwise an extremely successful mission.

Further variety was introduced into the work of the squadron during May. The CO, Maj W. A. McCloughry, undertook to use the Camel as a night-fighter, in unsuccessful attempts to bring down twin-engined Gotha bombers which were coming over whenever the weather was favourable. The use of the tricky little fighter in this role had been proved feasible by Home Defence pilots in England towards the end of 1917, but its use at night in France was still rare enough for McCloughry's missions to be worthy of note.

Also, enemy observation balloons soon began to appear on the victory lists of various pilots. Although in themselves defenceless, these large and outwardly easy targets were well protected by machine-gun nests on the ground as well as by fighter patrols which were usually in close proximity. To destroy one was no mean effort, and the effect of its loss on the enemy artillery well justified its inclusion as a victory. Except for unpremeditated attempts such as the ill-fated attack by Randell, no serious work of this nature had come the way of No 4, but the increasing use of these airborne observation posts in their area resulted in some extensive activity against them. Cobby was the first to add one to his score on 21st May, and this sparked off a succession of such victories by various pilots. Cobby and

King George V talks to Lt R. King and Capt A. H. Cobby of No 4 Squadron in August 1918. At His Majesty's right elbow is Maj W. A. McCloughrey, CO of the squadron, and in the right foreground, Gen Birdwood. Paraded officers are (from left) Lts W. C. Trescowthick, R. G. Smallwood, L. Bayer, R. R. B. Bennett, D. C. Carter, T. R. Edols, M. H. Eddie and L. T. E. Taplin.

Capt E. F. Pflaum, an original No 4 Squadron pilot, with his white Camel, serial C123, while instructing at Leighterton. The distinctive color was necessary to enable pupils to distinguish the aircraft easily during practice combat. Right. Sopwith Camel C106 after a bad landing at Leighterton. Front-line Camels were usually rigged tail-heavy so that any relaxation on the control column resulted in a tactically advantageous steep climb.

the New Zealander H. G. Watson seemed to be particularly infected by the occupational disease known as "balloon fever". Fortunately, unlike many other notable exponents of the art, neither of them fell victim to its hazards. However, there were casualties, two occurring simultaneously on 22nd May when Lts G. Nowland and A. Finnie collided and were killed during a balloon attack (before he suffered the same fate, Nowland was the last of the three pilots who replaced the January collision victims). Several more pilots were lost, either killed or captured, but the enemy suffered to a much greater extent during the constant operations of May and June.

In the second half of 1918 No 4 joined forces with their fellow SE5a unit. Operating from the same aerodrome at Reclinghem, the two squadrons maintained an attacking force which sought the enemy from ground level to 20,000 feet, and from the front lines to any point within the endurance of their aircraft. The Camel squadron, which was more fully occupied with the lower regions, consequently had a greater share of the air action. Apart from normal offensive patrols, their ground attack work often made them the target for hopeful enemy aircraft, and considerable numbers of observation machines also operated in the air space controlled by the Camels. Thus, No 4 began to realise its stated ambition to become the highest-scoring Australian air unit (the fact that they eventually achieved this ambition does not detract from the work of No 2, whose protective umbrella allowed the Camels to concentrate on the business in hand, while relatively free from interference from above). The efficiency of the Camel has already been mentioned, and a definite superiority was established over most of the types encountered. The pilots had to use every ounce of their skill against the increasing number of Fokker DVII's, but still achieved much success at the lower altitudes where the two types were more evenly matched. A fight between equal numbers of these two types on the last day of July resulted in two Fokkers being destroyed and two more sent down out of control, the damage being done by three comparatively new members of the squadron, Lts L. T. E. Taplin, T. R. Edols and T. C. R. Baker. Taplin, the most recent replacement, had only joined at the beginning of the month, but already the value of his experience with No 1 was making him a valuable asset.

With the opening of the final phase of the war in August, operations against the enemy were stepped up, but the considerable successes were offset by a series of casualties. One of the new pilots, Lt L. Wharton, found himself separated from his companions on the 11th but, undaunted, took on two LVG two-seaters. After sending one down, the Australian must have received the impression that the whole German Air Service was at his heels when half a dozen fighters shrieked down seeking revenge. With his Camel badly damaged and petrol pouring from holes in the tank behind his seat, Wharton sought safety in the clouds below. Unfortunately, after emerging from their shelter, he was wounded by anti-aircraft fire and crashed on the British side

Lt G. Jones (now Air Marshal Sir George Jones, KBE, CB, DFC) with a Sopwith Camel at Leighterton during a period of convalescence from a wound received in March 1918. He later returned to No 4 Squadron and was a flight commander when the war ended.

In October 1918 No 4 Squadron received the Sopwith Snipe. E8198 of "C" Flight displays its pugnacious lines at Bickendorf aerodrome, Cologne, during the squadron's post-war occupation duties. Right. Snipes of No 4 Squadron at Bickendorf. "F" of Lt E. A. Collins shows the common discoloration of the white bar caused by the pilot's oil-soaked boot when alighting after flight.

of the lines, later being invalided back to England. A few days later Lt S. Howard, who had joined the squadron with Wharton, was wounded during an attack on ground positions, along with Lt R. G. Smallwood, and they also went to hospital. Lt Edols followed them after being hit on the 25th, but he was more fortunate than Wharton's replacement, Lt S. P. Keay, who was killed on the 22nd as a result of another of the bad collisions which seemed to plague No 4. Lt E. A. Cato, the other victim of the mishap, somehow survived his 120 mph crash and sustained only minor injuries! During this period, the only fatality due to enemy action occurred on the 17th, when Lt E. P. E. McCleery was killed during the big combined raid on Lomme aerodrome. His death was the first in action that the squadron had suffered in over two months.

September saw No 4 suffer an even worse blow, when four members of a five-man patrol were shot down on the 5th by an overwhelming number of German fighters. Lts M. H. Eddie, D. C. Carter and A. H. Lockley were killed and Taplin was wounded and captured, but not before he bested several of his attackers. The lost pilots were well avenged on the following day when seven Fokker DVII's made the mistake of attacking an innocent-looking patrol of six Camels without taking the elementary precaution of checking the sky above. Six aircraft from No 2 had been waiting for just such an opportunity, and between them the Australians left very little of the shattered enemy formation to find its way home.

From now until the last few weeks of the war the enemy air opposition was spasmodic but no less fierce, and the pilots of No 4 continued to take their toll. In October the squadron was re-equipped with the new and more efficient Sopwith Snipe, which meant that they could at last take the fight up to the high-flying masses of German aircraft which were now beginning to appear.

This they did to very good effect, but with such fierce fighting casualties were unfortunately high. Lt P. J. Sims was killed and Lt M. J. Kilsby was taken prisoner on successive days at the end of October and, a week before the Armistice came into effect, the squadron suffered the loss of five pilots in one day. Lts E. J. Goodson and C. W. Rhodes were taken prisoner in the morning and in the afternoon Capt Baker, the leader of "B" Flight, and Lts A. J. Palliser and P. W. Symons were all killed during one short but particularly torrid engagement. The large number of enemy aircraft destroyed by the squadron during the previous week was little compensation for their loss.

That black day of 4th November 1918 marked the last serious combat by the pilots of No 4 and they completed their war service by reversing roles and acting as top cover while the SE's of No 2 pounded the retreating German infantry. After hostilities finished, the squadron was ordered into

Snipe cockpit. Note the padding on the end of the gun breeches to protect the pilot in a crash.

Officers of No 4 Squadron AFC with a Sopwith Snipe at Bickendorf aerodrome, Cologne.

Germany as part of the occupation forces and remained at Cologne until disbandment in February 1919, finally leaving England in May with the remainder of the AFC personnel for Australia.

Once back home, many of these men found it hard to adjust themselves to peacetime conditions; indeed quite a few had never known a normal adult civilian way of life. It was only natural that such men would seek some form of employment with flying, and what better place to start than a wide-open continent with a limitless expanse of sky? Many branched out on their own, heedless of the unforeseen difficulties associated with their new ventures but, whatever the final result, they showed the same zest and spirit of adventure as they displayed in their service flying. These were the visionaries who laid the foundations for Australia's great civil airline network. And in the same way yet another group looked ahead to the important part that the aeroplane would play in the future defence of Australia, and struggled with equal tenacity to gain recognition for and the development of the Service that finally emerged in 1921 as the Royal Australian Air Force.

Original members of No 4 Squadron pose in front of a German Gotha GVII at Bickendorf in December 1918.

"The Grog Wagon" was the nickname for this German Rumpler used by No 4 as a communications and supply aircraft in Germany. Capt J. W. Wright is the pilot with Capt E. F. Pflaum as passenger. Right. Lt E. A. Cato peers uncomfortably from the nose compartment of a German AEG GIVK at Bickendorf.

Capt Pflaum's white Avro 504K at Leighterton. The prominent display of his nickname "Tab" was the result of an early morning "raid" by the instructors on the sleeping quarters of the CO. Unable to positively identify the "attacking" aircraft, the latter ordered this step to discourage repetition and promote sounder sleep! Right. The kangaroo and joey mascot which alternately graced Capt Pflaum's Avro and Camel at Leighterton.

Low-level demonstration by an Australian Snipe in Germany. German pilots were very impressed with the qualities of the machine and expressed relief that they had no further need to meet it in combat. Right. The final mission. AFC pilots group together on the deck of the "Kaiser-I-Hind" in May 1919, homeward-bound to Australia and demobilisation.